Michael J. Spear

Life Science
CATHOLIC HERITAGE EDITION

Student Workbook

Catholic Heritage Curricula
1-800-490-7713   www.chcweb.com

Credits

Interior Design: Lauren Woodrow

Image credits: Line art illustrations: Michael J. Spear; cover: © Ruta Saulyte-Laurinaviciene/iStock/Thinkstock.com, © Fuse/Fuse/Thinkstock.com, © Fuse/Fuse/Thinkstock.com, © michaeljung/Shutterstock.com; pg. i: © Ruta Saulyte-Laurinaviciene/iStock/Thinkstock.com; pg. 1: © Martin Shields/Science Source; pg. 2: © alexyndr/Shutterstock.com; pg. 24: © Tshooter/Shutterstock.com; pg. 112: Special thanks to Kada at Kada's Garden, www.kadasgarden.com; pg. 113: Special thanks to Dr. Thomas Ombrello; pg. 237: © Grasko/Shutterstock.com; pgs. 274–276: Special thanks to the New York State Conservationist magazine; back cover: © oksana.perkins/Shutterstock.com, © ephotographer/Shutterstock.com, © oorka/Shutterstock.com, © Poznukhov Yuriy/Shutterstock.com, © markrhiggins/Shutterstock.com, © Valerie Potapova/Shutterstock.com.

ISBN: 978-0-9913264-1-9

Original edition © 1998 Helene Spear

Catholic Heritage Edition © 2014 Theresa A. Johnson

This book is under copyright. All rights reserved. No part of this book may be reproduced in any form by any means—electronic, mechanical, or graphic—without prior written permission. Thank you for honoring copyright law.

Printed by Sheridan Books Inc., Chelsea, Michigan
April 2016   Print code: 383147

For more information:
Catholic Heritage Curricula
1-800-490-7713
www.chcweb.com

Contents

To the Teacher, *1*

Chapter 1, *5*
  Formal Lab: Measuring the Strength of Different Seedlings

Chapter 2, *17*
  Diagram: Centimeter Ruler
  The Compound Microscope
  Microscope: Paper

Chapter 3, *28*
  Leaf or Insect Collection
  Microscope: Butterfly Wing

Chapter 4, *31*
  Diagram: Bohr Models
  Microscope: Salt Crystals

Chapter 5, *39*
  Experiment: Learning about Osmosis
  Microscope: Onion Cells
  Formal Lab: Osmosis

Chapter 6, *51*
  Formal Lab: Comparing the Bacteria on Different Surfaces

Chapter 7, *63*
  Diagram: Paramecium
  Microscope: Protists

Chapter 8, *73*
  Diagram: Typical Mushroom
  Experiment: Spore Prints
  Microscope: Mushroom Spores
  Microscope: Yeast Cells

Chapter 9, *83*
  Diagram: Fern
  Microscope: Green Algae
  Diagram: Plant Cell

Chapter 10, *95*
  Diagram: Root
  Diagram: Leaf
  Microscope: Xylem and Phloem in a Celery Stem
  Microscope: Parts of a Plant
  Microscope: Leaf Stomata

Chapter 11, *105*
  Diagram: Complete Flower
  Diagram: Five Stages of Mitosis
  Experiment: Grafting Cacti
  Microscope: Cactus Spines

Chapter 12, *115*
  Diagram: Metamorphosis
  Diagram: Generalized Insect
  Experiment: Animal Dissection

 Detailed, daily lesson plans are available at *www.chcweb.com*.

Chapter 13, *125*
    Diagram: Animal Cell
    Microscope: Feather

Test: Midway Review, *135*

Animal Classification Research Paper, *141*
    Tips for Writing a Research Paper

Chapter 14, *147*
    Formal Lab: Yeast and Sugar

Chapter 15, *161*
    Experiment: Test for Starch
    Experiment: Test for Simple Sugars

Chapter 16, *169*
    Diagram: Human Body
    Sanctity of Human Life Report
    Microscope: Cheek Cells

Chapter 17, *177*
    Diagram: Knee Joint
    Diagram: Types of Muscle Cells
    Experiment: "Rubberize" a Bone

Chapter 18, *185*
    Diagram: Digestive System
    Formal Lab: Chemical Digestion of Starch

Chapter 19, *197*
    Diagram: Heart
    Microscope: Blood Cells

Chapter 20, *205*
    Diagram: Respiratory System

Chapter 21, *215*
    Diagram: Kidney
    Diagram: Skin
    Microscope: Hair

Chapter 22, *223*
    Research on a Dwarf or Giant

Chapter 23, *229*
    Diagram: Neuron
    Formal Lab: Distance between Nerves

Chapter 24, *243*

Fight against Disease Research Paper, *254*

Chapter 25, *257*
    Formal Lab: Sensory Response Times

Chapter 26, *267*
    Diagram: Food Web
    Diagram: Nitrogen Cycle

Test: Final Review, *275*

A Miniature Ecosystem, *283*

Answer Key, *287*

To the Teacher:

As a Catholic husband and a father of seven, I am interested in how the Faith of the Ages may be transmitted to my children, not as a set of rules, but as the very foundation of life! I feel strongly that a deep interest in science will lead to a stronger faith because the genuine scientist is a seeker of truth. Science teaches that there are objective answers that are correct while other answers are definitely incorrect. Science teaches that there is order in the natural world of which we are a part. All this can lead to an understanding that there is right and wrong and that there is an Almighty who is indeed Master of the Universe.

This leap from science to faith may be accomplished if the individual has learned to put self aside and to look to the welfare of others. So develop, most importantly, the individual first, then with sound scientific thinking the mind will come to the Faith that every soul seeks.

The text of *Life Science* is written with the idea that your child will learn what is expected of him or her. Plenty is expected and plenty can be learned. The course covers only the basic ideas of most topics. It does so clearly with ample review questions to reinforce the essential material.

Your students will be best served by *Life Science* if you use my two T's, M's, and E's:

- Teach the Top
- Measure the Middle
- Emphasize the Essentials

Teach the Top means to teach the very highest level possible with your student, and then go just a little higher. Expect the very most in this course. *Measure the Middle* reminds you to test and evaluate the student based on the information that appears most often and is given clear emphasis. Require responses for information that will appear most often in later courses and in life. Don't quiz the student on minute details of fact on one hand; on the other, don't expect him to present essays on the great ideas of life science. *Emphasize the Essentials* requires memorization and practice, lots of vocabulary, and more review than usual. Buttress the basics so that your student will never lose them.

Many of the review questions instruct the student to write complete sentences and definitions. Do not cut corners. Each statement is designed to reinforce a piece of information. Permanent learning requires more than A-B-C or 1-2-3 answers. Full sentences and complete ideas are remembered better when they are written out completely and neatly.

(continued on next page)

Difficult vocabulary is not to be avoided but learned. As the student learns scientific terms in his keywords lists, he should practice them orally and in writing so he becomes confident in using them. Scientific names of species can be easily handled by reading only their initials when the common name is also in the sentence.

Science and faith are fully compatible. We must learn all we can and then put our knowledge and our lives at the discretion of the "owner of the vineyard." Let us always strive to follow the advice of Saint Paul that we should "prove all things; hold fast that which is good" (1 Thessalonians 5:21).

—Michael J. Spear

Experiments and Labs

This workbook includes a wealth of experiments and microscope assignments to demonstrate and reinforce the scientific principles explained in the text. Experiment supplies may be collected and/or purchased ahead of time. See pages 3-4 for a complete list.

The seven **Formal Labs** assigned throughout the course provide the student with practice using the scientific method in a controlled experiment. Step-by-step instructions in the workbook guide the student in forming a hypothesis and testing it with a formal experiment that includes a control, an independent variable, and a dependent variable. The student then completes a formal laboratory report for each of the controlled experiments. These formal experiments and lab reports provide valuable experience to prepare the student for laboratory sciences at the high school and college level.

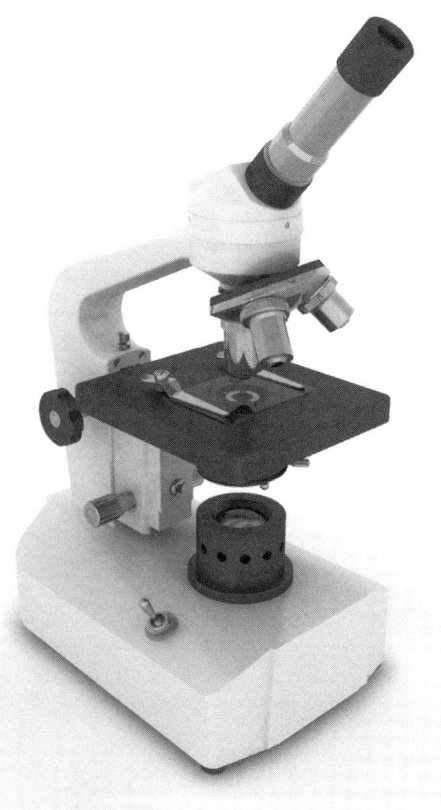

Microscope assignments are not essential to the course and may be omitted if the student does not have access to a microscope. Use of a microscope, however, can greatly enrich the study of biology both in junior high and high school, so the parent may find a microscope to be a worthwhile investment. Many microscopes come with prepared slides, which can be organized by chapter topic and brought out to view during the appropriate week.

Many of the microscope assignments include space for optional "Notes about specimen or procedure." Making such notes is optional, but the space can be useful for noting down tips for future reference, such as which samples were most impressive, etc. The student may also use the space to write a description of what he saw, such as, "I saw an amoeba eating a paramecium!"

⚠ Parental Supervision Advised: Although every effort has been made to ensure the safety of the experiments and labs within this workbook, parents are responsible for taking appropriate safety measures and supervising students. Catholic Heritage Curricula disclaims all responsibility for any injury or risk which is incurred as a result of the use of any of the material suggested in this course.

Using This Workbook

This workbook provides student-friendly exercises, research assignments, experiments, formal labs, keywords for memorization, diagramming assignments, tests, and a complete answer key to accompany *Life Science: Catholic Heritage Edition*.

Encourage your student to refer to the text to complete the **exercises**. The **tests** (Midway Review and Final Review) are designed to be given "closed-book." For each **keyword list**, direct student to fold the page on the dotted line so that the definitions are hidden beneath the fold. The student should quiz himself daily until he can explain the meaning of each term without looking at the definition. Frequent review of keywords from previous chapters is critical to doing well on the Midway Review and Final Review. The **Answer Key** (beginning on page 287) may be removed at the parent's discretion.

Two **research papers** are assigned in the course of this study, one on a particular animal and its classification and another on a scientist who contributed to the fight against disease. Detailed instructions for writing a research paper can be found on pages 143–146. These assignments provide valuable writing experience, especially if the student has never written a research paper before. The parent may wish to grant English credit as well as science credit for these papers, or may substitute them for English writing assignments.

This course is designed to be completed in one year, but it can also be taught over two years to allow more time to discuss and investigate topics of interest. Daily lesson plans to assist in teaching this course are available from Catholic Heritage Curricula (*www.chcweb.com*).

The **Check It Out!** web links listed throughout the workbook may be explored as time allows. Interactive links, extension activities, and video clips can make difficult concepts easier to grasp. Parental supervision is strongly advised when your student is visiting recommended website links. At the time of this printing these links contained helpful, age-appropriate information. Be advised that good links can change frequently, becoming inappropriate or no longer available. Because of this, please visit recommended web links beforehand and supervise student during his online visit as he explores the activity, video, or experiment.

Supply List

Most of the supplies listed on the following page are readily available to the homeschooling family, but a few items will need to be purchased from a science lab supply company such as *www.enasco.com* or *www.carolina.com*. Keep in mind that some supplies, such as small, potted cacti, may not be available in your area during the winter, so you may wish to obtain them at the beginning of the school year.

In Chapter 3 the student is instructed to make a leaf or insect collection. The supplies needed for this assignment are not included in the supply list, since they depend on whether the student is collecting leaves or insects, and also on how simple or complex you wish the collection to be. (For instance, a shoe box is sufficient to store a simple insect collection, but a wooden display case might be better if your student is planning to enter his insect collection in a competition.) Detailed instructions for the collection assignment can be found in Chapter 3 of the student text and page 29 of the workbook.

Supply List

CHAPTER 1
- rich soil
- seeds: sunflower, corn, bean, radish (Use fresh seeds for best results.)
- 20 light-weight, disposable plastic cups
- a few dozen tiny scraps of paper
- large tray or planter box, about 18" × 24" (A temporary box can be made from a shallow cardboard box lined with a large plastic trash bag.)
- 640 grams of change (for instance, 128 nickels)
- scissors

CHAPTER 2
- microscope, slide and slip
- small scraps of a variety of different types of paper (for instance, colored construction paper, laser paper, handmade paper, a napkin, newspaper)
- small piece of paper with printed words
- ruler to measure millimeters

CHAPTER 3
- microscope, slide and slip
- wing of a dead butterfly or moth
- eyedropper
- paintbrush or finger

CHAPTER 4
- microscope, slide and slip
- table salt
- small drinking glass
- eyedropper

CHAPTER 5
- drinking glass or mug
- raw egg in shell
- vinegar
- spoon
- microscope, slide and slip
- piece of raw onion (white or yellow)
- eyedropper
- tincture of iodine or Lugol's solution (optional)
- two potatoes
- ¼ cup salt
- plastic wrap
- paring knife
- cutting board
- two small bowls
- 1-cup measuring cup
- spoon

CHAPTER 6
- 125-ml bottle of liquid nutrient agar
- five disposable petri dishes
- household bleach
- permanent marker
- q-tips
- stove
- saucepan
- tape
- five ziplock plastic bags

CHAPTER 7
- microscope, slide and slip
- pond water or seawater
- jar or bucket
- eyedropper

CHAPTER 8
- two freshly picked, mature mushrooms (Their gills must be visible when they are picked.)
- one sheet of white paper
- one sheet of black paper
- microscope, slide and slip
- eyedropper
- active dry yeast
- sugar
- small bowl

CHAPTER 9
- microscope, slide and slip
- green algae
- transparent glass or jar
- tweezers
- eyedropper
- Lugol's solution or tincture of iodine

CHAPTER 10
- stalk of celery
- paring knife
- tall drinking glass
- food coloring (blue or red)
- spoon
- microscope, slide and slip
- eyedropper
- flower petal
- leaf
- root hair
- clear nail lacquer (Nail *polish* or *enamel* is too thin, and tends to stick to the leaf.)
- a sturdy leaf that is not covered with "hairs" (Leaves from a rose bush work well.)

CHAPTER 11
- two small cacti of different varieties (For best results, choose cacti with thick, barrel-like stems.)
- sharp paring knife
- string or rubber bands
- gardening gloves
- microscope, slide and slip
- spines from two or more types of cacti
- tweezers

CHAPTER 12
- preserved animal for dissection
- household cutlery or dissection kit

CHAPTER 13
- microscope, slide and slip
- feather
- eyedropper

CHAPTER 14
- active dry yeast
- sugar
- five identical, empty water bottles with tight-fitting caps
- five standard balloons of the same size
- five rubber bands
- permanent marker
- measuring spoons
- sheet of paper
- 15-in piece of string
- centimeter ruler

CHAPTER 15
- cracker
- piece of cheese
- slice of apple
- pat of butter
- peanut
- Lugol's solution or tincture of iodine
- eyedropper
- a small amount of: meat (cooked or raw), bread, milk, carrot (or another vegetable), apple (or another fruit), table sugar
- Benedict's solution
- saucepan
- one or more glass test tubes

CHAPTER 16
- microscope, slide and slip
- tincture of iodine or Lugol's solution
- toothpick
- eyedropper

CHAPTER 17
- cooked chicken bone
- jar or glass
- white distilled vinegar

CHAPTER 18
- cooked rice
- Benedict's solution
- at least one glass test tube
- saucepan
- knife
- spoon

CHAPTER 19
- microscope, slide and slip
- drop of blood

CHAPTER 21
- microscope, slide and slip
- different types of hair: straight, curly, brown, blonde, thin, thick, etc.

CHAPTER 23
- two paperclips
- centimeter ruler

CHAPTER 25
- centimeter ruler (at least one foot long; a meter stick is ideal)
- four friends or family members

MINIATURE ECOSYSTEM
- ten-gallon tank
- clean water (from stream, pond, or well)
- clean soil
- small sample of animal and plant life (including snails)

Chapter 1

Workbook 1.1

Sample

Use the glossary to define each of the keywords on the lines below, then fold the paper on the dotted line so the definitions are hidden beneath the fold. Quiz yourself on the keywords every day until you can explain the meaning of each term without looking at the definition.

ABSORPTION ....

BIOLOGY ....

CONTROL ....

DEPENDENT VARIABLE ....

DIGESTION ....

EXCRETION ....

GROWTH ....

HYPOTHESIS ....

INDEPENDENT VARIABLE ....

INGESTION ....

IRRITABILITY ....

METABOLISM ....

NUTRITION ....

ORGANISM ....

OXIDATION ....

PHOTOSYNTHESIS ....

REPRODUCTION ....

RESPIRATION ....

SYNTHESIS ....

TRANSPORT ....

Chapter 1

Workbook 1.2

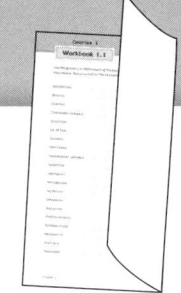

Sample

Learn the biology root words listed below. Quiz yourself on the keywords until you have memorized them.

Root word	Meaning	Examples
A OR AB	NOT	ABIOTIC, ANAEROBIC
ANTI	AGAINST	ANTIBODY, ANTITOXIN
AQUA	WATER	AQUARIUM, AQUATIC
AUDIO	SOUND OR HEARING	AUDITORY NERVE
BI	TWO	BINARY FISSION, BILATERAL, BICEPS
BIO	LIFE	BIOLOGY, BIOME, BIOSPHERE, BIOTIC
CARDIAC	REFERRING TO THE HEART	CARDIAC MUSCLE
CELLA	SMALL ROOM	CELL, CELLULAR
CENTI	100	CENTIMETER, CENTURY, CENTS
CHROMO	COLOR	CHROMOSOME
CYANO	DARK BLUE	CYANOBACTERIA
CYKLOS	CIRCLE	CYCLOSIS, CIRCULAR, CYCLE
DECI	TEN	DECIMETER, DECIMAL, DECADE
DERMA	SKIN	EPIDERMIS, MESODERM
DI	TWO	DIVIDE, DISSECT
EPI	UPON	EPIDERMIS
EX	OUT	EXIT, EXCRETION, ECTODERM
GASTRO	REFERRING TO STOMACH	GASTRIC JUICE, GASTROINTESTINAL
GEO	EARTH	GEOTROPISM, GEOGRAPHY
GLOTTIS	TONGUE	GLOTTIS, EPIGLOTTIS
GRADE	GRADUATION OR STEP	GRADUATIONS, CENTIGRADE
IN	IN	INHALE, INGEST, INSPIRATION
MACRO	BIG	MACRONUCLEUS
MARINE	OCEAN	MARINE, SUBMARINE
METER	TO MEASURE	METRIC, METER, THERMOMETER
MICRO	SMALL	MICRO-ORGANISM, MICROSCOPE
MILLI	1000	MILLIMETER, MILLILITER, MILE
MULTI	MANY	MULTICELLULAR
OLOGY	THE STUDY OF	BIOLOGY, ORNITHOLOGY
PHOTO	LIGHT	PHOTOSYNTHESIS, PHOTOTROPISM
PULMO	REFERRING TO THE LUNG	PULMONARY CIRCULATION
RE	TO DO AGAIN	REPRODUCE, REVIEW
SEMI	PARTLY	SEMIPERMEABLE
SOMA	BODY	CHROMOSOME
SUB	UNDER	SUBMARINE, SUBCUTANEOUS
SYNTHESIS	TO MAKE	SYNTHESIS, PHOTOSYNTHESIS
TRANS	ACROSS	TRANSPORT, TRAIN
VACCINUS	COW	VACCINE, VACCINATION
VIDEO	SEE	VISION, VIDEO

Workbook 1.3

A. Write the keyword from Chapter 1 that is most closely associated with each phrase. Keywords are capitalized in the body of the text.

1. After 21 days a chick hatches from an incubated egg. ....

2. Cows have flat molars to mash up the hay they chew. ....

3. Each cell of a pine tree combines oxygen with glucose to get energy.
   ....

4. A painted turtle eats an earthworm. ....

5. Glucose is being absorbed by the villi in the wall of your intestine.
   ....

6. You blink when a camera flash goes off. ....

7. A patient's heart must continue to beat if he or she is to remain alive.
   ....

8. The body breaks down the starch in French fries to form a simple sugar.
   ....

9. Plants use chlorophyll to capture light and make their own food.
   ....

10. You exhale carbon dioxide through your mouth and nasal passages.
    ....

B. Using a full sentence, explain why salt crystals, which show growth, one of the life functions, are not alive.

....
....
....
....

Chapter 1

Workbook 1.4

Write the name of one of the parts of a book that would be the BEST section to use to find the information.

1. Which part of this book will give you the chapter that discusses cells?

   ....

2. Which part will contain the best definition of metabolism? ....

3. In which section might there be a chart listing the uses of plants?

   ....

4. Where should a pupil look to read all about the human skeleton?

   ....

5. Which part is the best to use for the meaning of the term "fetus"?

   ....

6. In reading that the tallest man ever was Robert Wadlow, which section should be used to check on how old this information is?

   ....

7. Which section will give the author's name? ....

8. Which part gives an outline of the book? ....

9. Which part is a small dictionary? ....

10. Which section includes a chart of word roots for you to learn?

    ....

11. Which section is the main part of the book, with paragraphs, photos, and diagrams?

    ....

Workbook 1.5

A. Answer each question below with a neatly written, complete sentence:

1. Which parts of a book are found in *Life Science: Catholic Heritage Edition*?
   ....

2. On which page does the glossary begin?
   ....

3. How are the words in the glossary arranged?
   ....

B. Define the parts of a book.

1. Title page ....

2. Contents ....

3. Body ....

4. Appendix ....

5. Glossary ....

1.6 Formal Lab #1

Complete the experiment below, following the instructions provided. Fill in any blanks as you come to them. Use complete sentences to answer the questions at the end.

Name: .... **Date:** ....

I. Title: Measuring the Strength of Different Seedlings

II. Purpose: When seeds sprout, the seedlings have to be strong enough to push through the dirt as they grow towards the sunlight. Many seedlings are even strong enough to push their way between gravel and through cracks in asphalt and cement.

In this experiment we will measure the strengths of sunflower, corn, bean, and radish seedlings, and attempt to determine which species possesses the most "lifting power."

My hypothesis is that the .... **[sunflower/corn/bean/radish] seedlings will be able to lift the most weight. I think this because** ....

....

....

....

III. Materials: The materials required for this experiment are rich soil, sunflower seeds, corn seeds, bean seeds, radish seeds, water, 20 light-weight, disposable plastic cups, and a few dozen tiny scraps of paper. (Use fresh seeds for best results.)

IV. Apparatus: The equipment required for this experiment is a large tray or planter box (about 18" × 24"), 640 grams of change (for instance, 128 nickels), and scissors. (A temporary planter box can be made from a shallow cardboard box lined with a large plastic trash bag.)

V. Procedure:

1. Cut the top halves off of 20 plastic cups, leaving the sides about 1½ inches high.
2. Prepare the planter box by filling it with the rich soil. Be sure to break up any lumps or clods of dirt, and moisten the soil thoroughly before filling the planter box. Select five seeds from each seed package, being careful not to choose any that look broken or unhealthy.
3. Plant four rows of five seeds each in the planter box, one row for each type of seed. Plant the seeds about three inches apart, and only plant one seed in each space. Follow the directions on the package for how deep to plant the seed. As you plant each seed, set a tiny scrap of

paper directly above the seed on top of the soil. This will help you keep track of where you planted the seed.

Mass of U.S. Coins
Penny = 2.5 g
Nickel = 5.0 g
Dime = 2.27 g
Quarter = 5.67 g

4. Make a note of which seeds you plant in each row, and then place a plastic cup directly over each seed. Leave the first cup in each row empty. These cups will be our control. Fill the other cups with coins so that the second cup in each row contains 10 grams, the next cup in each row contains 25 grams, the fourth cup contains 50 grams, and the fifth cup in each row contains 75 grams of change. One nickel weighs five grams, so if you are using nickels, you will put two nickels in the second cup of each row, five nickels in the next cup, 10 in the fourth cup, and 15 in the fifth cup.

5. Put the planter box in a warm, sunny place. Check the planter box regularly, watering it as needed to keep the ground moist, but not soggy. When you water, be sure to keep the cups centered over the scraps of paper that mark where the seeds were planted.

6. When the seeds begin to germinate, keep track of which seeds are able to lift the cups by filling in Figure 1. Write "yes" in the chart for the seedlings that lift their cup, and write "no" for the seedlings that sprout but are not able to lift their cups.

7. If one of the seeds does not germinate within the time stated on the seed package, or if most of the other seeds of that species have already germinated, remove its cup and plant a fresh seed in its place. Put the cup back over the seed. When the new seed has germinated, record whether it can lift its cup.

8. When all the seedlings have germinated, use the data you have recorded in Figure 1 to fill out Figure 2. Then fill in vertical bars in Figure 3 to indicate the heaviest weight that each species was able to lift. For instance, if the heaviest weight lifted by the corn seedlings was 50 grams, the bar for corn will reach to the horizontal line that is labeled "50" on the left side of the graph.

VI. Data:

	Sunflower	Corn	Bean	Radish
75 g				
50 g				
25 g				
10 g				
Control				

FIGURE 1. WEIGHT LIFTED BY EACH PLANT

	Heaviest weight
Sunflower	
Corn	
Bean	
Radish	

FIGURE 2. HEAVIEST WEIGHT LIFTED

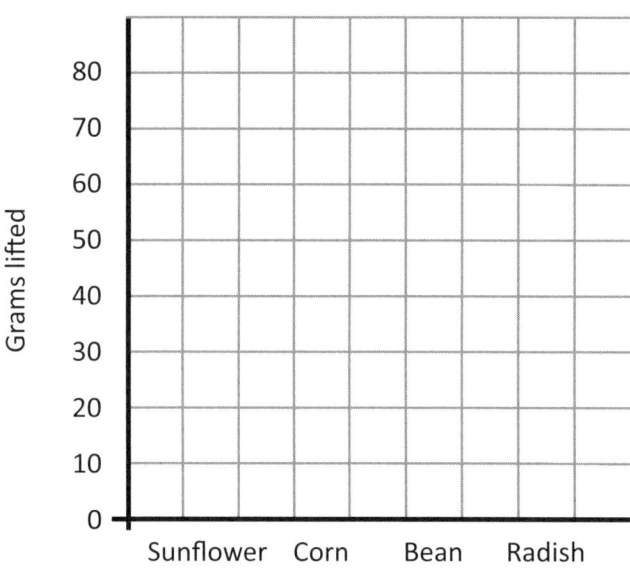

FIGURE 3. STRENGTH OF DIFFERENT TYPES OF SEEDLINGS

VII. Questions:

1. Did one species of plant have more lifting power than the others? If so, which was strongest? Which was weakest?

   ....
   ....
   ....

2. The plants grown under the empty cups were the **control** in this experiment. The **independent variable** was the different types of plants that we used. What was the **dependent variable**? In other words, what variations in the results were caused by using different species of plants?

   ....
   ....

3. What happened to seedlings which were unable to lift their cups? Were any of them able to reach the sun by a different method?

   ....
   ....

VIII. Conclusion: This experiment showed that the .... [sunflower/corn/bean/radish] seedlings were able to lift the most weight.

Thus, my hypothesis was .... **[correct/incorrect].**

Extra credit:

What do the results of the experiment indicate about what gives seeds their strength? For example, were the largest or the smallest seeds stronger? Were seeds of a certain type, such as monocots vs. dicots, stronger than others of approximately the same size? You may wish to refer to the chart in Chapter 9 in the text (page 81) which compares monocots and dicots.

....
....
....
....
....
....

Chapter 1

Chapter 2

Workbook 2.1

Sample

Use the glossary to define each of the keywords on the lines below, then fold the paper on the dotted line so the definitions are hidden beneath the fold. Quiz yourself on the keywords every day until you can explain the meaning of each term without looking at the definition.

- Area ....
- Centi ....
- Centigrade ....
- Deci ....
- Deka ....
- Graduation ....
- Gram ....
- Hecto ....
- Kilo ....
- Liter ....
- Mass ....
- Mega ....
- Meter ....
- Meter Stick ....
- Metric System ....
- Micro ....
- Micron ....
- Milli ....
- Unit ....
- Volume ....

Workbook 2.2

Diagram and label a centimeter ruler in the space below. Refer to Figure 2.2 in the text. You may use a pencil to draw the diagram, but use a pen to write the labels. Use a ruler for any straight lines, and don't forget to write a title.

Title: ....

CHECK IT OUT!   http://learn.genetics.utah.edu/content/begin/cells/scale/

Workbook 2.3

A. List the metric prefixes from the largest we've studied to the smallest, including the blank spaces (see Figure 2.3 in the text). Write your list from left to right across the page.

....

B. Write the meaning of each abbreviation.

1. m: ....
2. mm: ....
3. ml: ....
4. dal: ....
5. cm^2: ....
6. l: ....
7. cm: ....
8. cg: ....
9. Mg: ....
10. cm^3: ....
11. g: ....
12. dm: ....
13. km: ....
14. µm: ....
15. cc: ....

C. Fill in the blanks with the correct number.

1. 1000 mm = ... m
2. 1000 mm = ... cm
3. 348 cm = ... m
4. 5 kg = ... g
5. 2.61 dl = ... cl
6. 0.01 cm = ... m
7. 0.01 mm = ... m
8. 0.056 m = ... km
9. 0.004 Mg = ... g
10. 0.025 l = ... ml
11. 2 hg = ... dag
12. 2 hl = ... dal
13. 2 hm = ... dam
14. ml = ... cc
15. 4,500 cc = ... l

Workbook 2.4

A. *Refer to Figure 1 to determine the quantities requested in numbers 1–4 below. Be sure to include units!*

1. What is the area of the side marked "HANDLE WITH CARE"? ....

2. What is the area of the side marked "EGGS"? ....

3. What is the area of the side marked "FRAGILE"? ....

4. What is the volume of the box in Figure 1? ....

B. *Refer to Figure 2 to determine the quantities requested in numbers 1–4 below.*

1. What is the area in square centimeters of the shaded side of the block? ....

2. What is the area of the white top of the block? ....

3. What is the volume, in cubic centimeters, of the block? ....

4. How many milliliters of water would occupy the same volume as the block? ....

C. *If the volume of a room is 36 cubic meters, what is its volume in cubic centimeters? Hint: draw and label a sketch of the room's possible dimensions!* ....

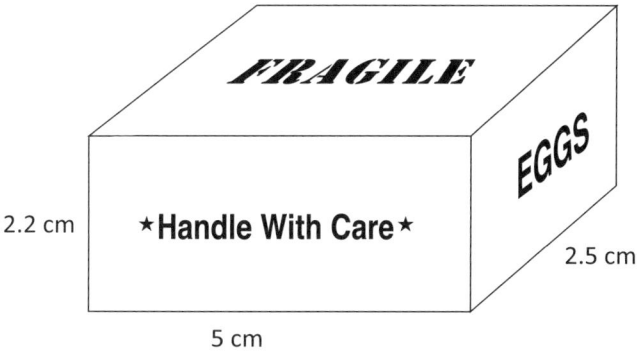

FIGURE 1. HUMMINGBIRD EGG BOX
This miniature box could hold 15 eggs of the Helena's Hummingbird, with room left over!

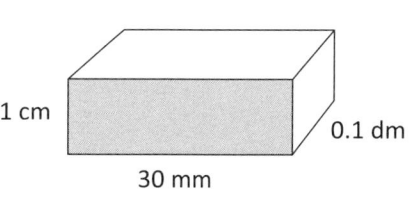

FIGURE 2. A BLOCK OF WOOD

Chapter 2

21

Workbook 2.5

Answer each question with a complete sentence.

1. How many cubic centimeters are equal to one milliliter?

   ....
   ....

2. What is the formula for rectangular area?

   ....
   ....

3. What is the formula for rectangular volume?

   ....
   ....

4. Upon what number is the metric system based?

   ....
   ....

5. What must be included with the magnitude (i.e. number) of a measurement?

   ....
   ....

6. What is the prefix that means 1000 times a unit?

   ....
   ....

7. Which prefix means 1/100 of a unit?

   ....
   ....

8. Which is larger: one decimeter (1 dm) or one dekameter (1 dam)?

   ....
   ....

9. Which of these is smallest in value: hecto, micro, milli, centi?

   ....

   ....

10. Give the root meaning of the prefix "*centi*."

    ....

    ....

11. Name two words that correctly contain the word root "*centi*."

    ....

    ....

Chapter 2

Workbook 2.6

The Compound Microscope

Imagine the excitement of Anton van Leeuwenhoek (LAY-wuhn-hook) when he looked into his first microscope and saw "wee beasties" so small that there are more of them on and in your body right now than there are people in the world! With a compound microscope, you will be able to see for yourself the tiny world of microbes (microscopic organisms).

The first microscopes from the late 1600s could only make things look approximately 150 to 200 times bigger. Today, life scientists often use the compound microscope to see things as small as 0.5 micron (called micrometer, μm). You can determine the power of your microscope by multiplying the strength of the eyepiece lens (or ocular lens) by the strength of the objective lens. For example, if the eyepiece is marked "10 X" and the objective is labeled "30 X," then the magnifying power of the scope is 10 × 30 or 300 times.

We must know the parts of a microscope if we are to use it correctly. Study Figure 1. This instrument is called compound because it has more than one lens. The microscope is carried by using one hand to hold the arm and the other to hold the base. Be sure to put the scope on a flat surface. When someone else wishes to see whatever you have in view on the stage, leave the microscope in one place and move yourself out of the way. To prevent damage to your eyes, never use direct sunlight with the mirror. Carefully following the instructions on page 25 will prevent damage to your microscope.

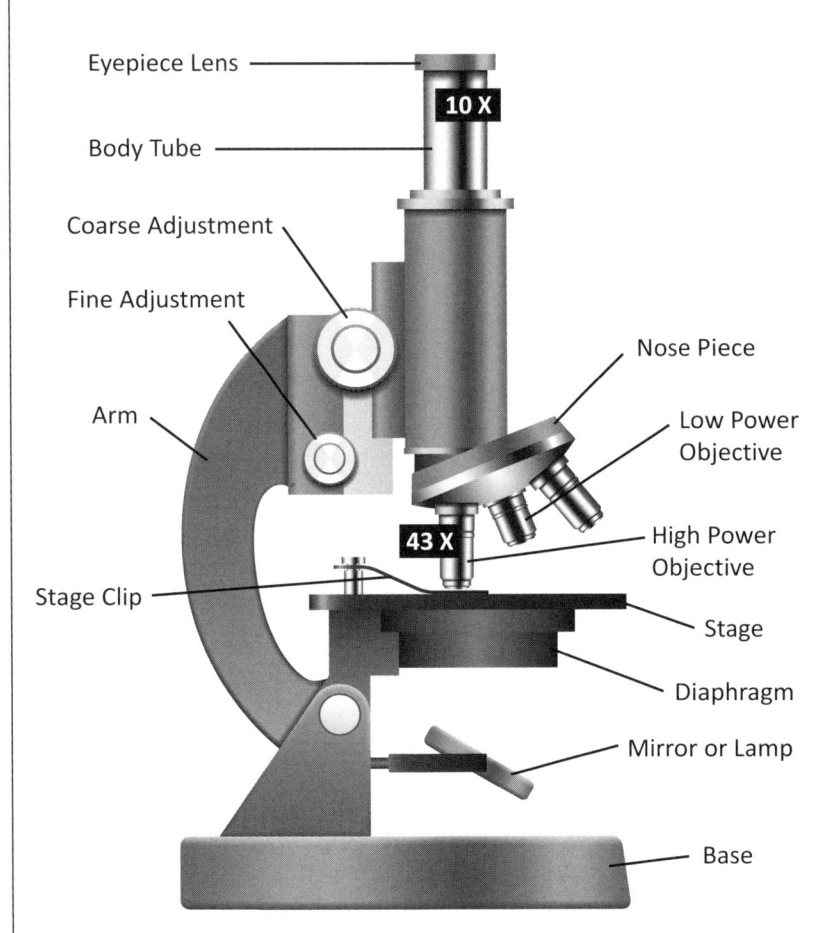

FIGURE 1. A COMPOUND MICROSCOPE
Clean the lenses with lens paper only.

Preparing a Wet Slide

Place your sample on the surface of a clean microscope slide. Use an eyedropper to add one drop of water to the sample.

Gently lower a cover slip onto the slide. To avoid creating air bubbles between the slide and the cover slip, lower the cover slip onto the slide at an angle, allowing one edge to touch the drop of water first.

Allow the liquid to spread between the cover slip and the slide.

Viewing a Prepared Slide

Turn on the microscope light or adjust the mirror. Make sure you are using the low power objective lens. Place the prepared slide onto the microscope stage, and use the stage clip to keep it centered beneath the lens.

Turn the coarse adjustment knob until the objective lens is just above the cover slip. View the microscope from the side (not through the eyepiece) while using the coarse adjustment, and be careful not to bring the objective lens too close. Otherwise, you will risk cracking the microscope slide and damaging the objective lens.

When the low power objective lens is just above the cover slip, look through the eyepiece. Turn the coarse adjustment in the opposite direction (increasing the distance between the objective lens and the slide) to bring the image into focus.

When you start to see an image through the eyepiece, use the fine adjustment knob to bring the image into focus. Use the diaphragm to increase or decrease the intensity of the light.

To view the slide on high power, turn the nose piece to switch the low power objective lens for the high power objective lens. Since the image was already in focus on low power, you will only need to use the fine adjustment to bring the image into focus. Never use the coarse adjustment when on high power or you may damage the objective lens by touching it to the slide!

When it is time to clean up, wash and dry your slide and cover slip. Return the microscope to low power, turn the microscope light off, and cover the microscope with its dust cover. Note that different microscopes may work slightly differently. For instructions that are specific to your model, refer to the user's manual that came with your microscope.

Figure 2 on the next page will help you understand what you are seeing in the microscope. Study it carefully and then test your knowledge with the quiz.

You can use a standard digital camera to photograph what you see through the microscope! When you have brought the image into focus, position the camera so that it is "looking" through the eyepiece. Take a photo just as you would normally, but do not use a flash. If you are not pleased with the quality of the photo, try adjusting the microscope diaphragm and the settings on the camera.

FIGURE 2. LOOKING INTO THE EYEPIECE

DARK

If you cannot see anything through the microscope, you should adjust the mirror or diaphragm to allow more light to reach the slide. If the view is too bright, use the diaphragm to reduce the light.

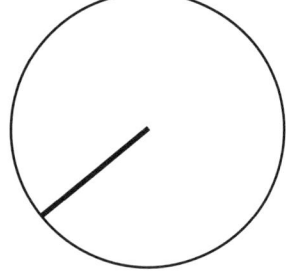
POINTER

The pointer can be used to point to a particular part of the specimen. The pointer is not movable, so you must move your slide to position whatever you want someone else to see at the tip of the pointer.

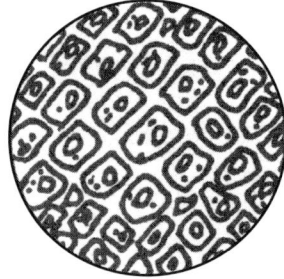
LOW POWER

The low power objective lens allows you to see more area but with less detail. You may use either the course or fine adjustment knob when using the low power objective lens.

HIGH POWER

Through the high power objective lens you will see more detail but less area. You must remember to use only the fine adjustment knob when on high power to avoid damaging the lens.

Quiz Yourself!

1. Determine the magnifying power of the microscope shown in Figure 1.

2. Should you use the coarse adjustment knob with the microscope set up as shown in Figure 1?

3. Bright light will "scare" many microbes away. Which part of a microscope allows you to reduce the light?

4. If you see only part of a cell under the microscope but want to see it all, should you use a higher power objective lens or a lower power objective lens to see more area?

ANSWERS:

1. 10 × 43 = 430 X (Eyepiece × Objective = Power)
2. Use only the fine adjustment when on high power.
3. The diaphragm is used to adjust the light.
4. A low power objective lens allows a larger field of view but less detail.

Workbook 2.7

Microscope: Paper

Supplies:
- microscope
- microscope slide and cover slip
- small scraps of a variety of different types of paper (for instance, colored construction paper, laser paper, handmade paper, a napkin, newspaper)
- a small piece of paper with printed words
- ruler to measure millimeters

1. Tear off a small piece of ordinary paper. Place it on a slide and cover with a cover slip. Do not add water. Follow the instructions on page 25 to view the prepared slide through the microscope.

2. Repeat with different types of paper. Be sure to view the surface of the paper as well as the torn edges.

3. Tear a small piece off the sheet of paper with printed words. Select a single letter and measure it in millimeters with a ruler.

4. Now view the letter under the microscope on low power. Using the size of the letter in millimeters, estimate the diameter of the area you can see through the eyepiece. Turn the microscope to high power and estimate the diameter of the area you can see through the eyepiece. Record your results below. Don't forget to include units.

Diameter of the area visible on low power:

....

Diameter of the area visible on high power:

....

Workbook 3.1

Use complete sentences to answer each question below.

1. List at least three different reasons for making a collection.

   ....

   ....

   ....

   ....

   ....

2. Name three kinds of plants that you should not collect.

   ....

   ....

   ....

3. How many leaf specimens should be taken from any one branch?

   ....

   ....

   ....

4. What plants require permission before taking a leaf?

   ....

   ....

   ....

5. Name a kind of "bug" that is not an insect.

   ....

   ....

   ....

6. List two kinds of insects that require caution when collected.

   ....

   ....

   ....

7. What four pieces of data should be included with every specimen?

   ....

   ....

   ....

   ....

   ....

8. Describe a type of container that should not be used to collect insects.

   ....

   ....

Leaf or Insect Collection

Take a few days to make your own life science collection! Begin by deciding whether you want to collect leaves or insects. Follow the instructions provided in Chapter 3 of the student text. The online resources below may also be helpful.

Leaf Collection and Identification

   http://texastreeid.tamu.edu/content/leafCollectingSafety/

   http://www.oplin.org/tree/

   http://dendro.cnre.vt.edu/dendrology/factsheets.cfm

Insect Collection and Identification

   http://extension.entm.purdue.edu

   http://www.extension.umn.edu/youth/mn4-H/projects/environment/entomology/collecting-and-preserving-insects/

   http://entnemdept.ufl.edu/bug_club/ent-events/collecting101.shtml

Workbook 3.2

Microscope: Butterfly Wing

Supplies:
- microscope
- microscope slide and cover slip
- wing of a dead butterfly or moth
- eyedropper
- water
- paintbrush or finger

1. Lay a small, flat portion of the butterfly wing on a microscope slide, being careful not to brush off the scales.

2. Place a drop of water on the wing piece, then lower a cover slip over the slide. Follow the instructions on page 25 to view the prepared slide through the microscope on low power and on high power. *Tip: If you find it difficult to focus the microscope, try adjusting the diaphragm to decrease the intensity of the light.*

3. Using a small paintbrush or your finger, take some scales off the butterfly wing and brush them onto a microscope slide. Gently lower a cover slip onto the slide. You do not need to add water. Follow the instructions on page 25 to view the prepared slide through the microscope on low power and on high power.

4. In the space below, draw the scales as you see them through the microscope. Be sure to label your drawing and record the magnification of the microscope. (See page 24 for instructions for determining the magnification of your microscope.)

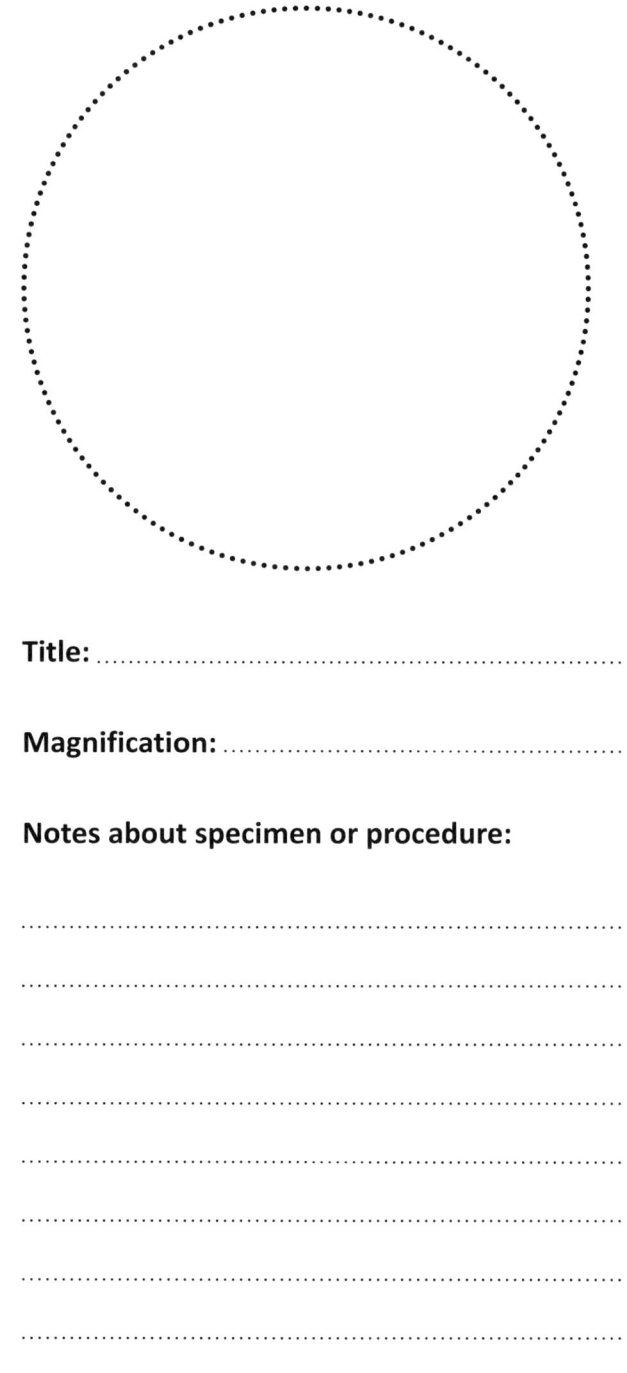

Title: ....

Magnification: ....

Notes about specimen or procedure:

....
....
....
....
....
....
....
....

Extra credit: Estimate the size of a single scale using your calculations from Workbook 2.6!

Chapter 4

Workbook 4.1

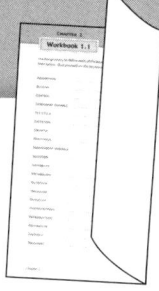

Sample

Use the glossary to define each of the keywords on the lines below. Quiz yourself on the keywords every day until you can explain the meaning of each term without looking at the definition.

Don't forget to review your keywords from Chapters 1-3! Frequent review of keywords from previous chapters is critical to doing well on the Midway Review and Final Review.

ATOM ....

ATOMIC NUMBER ....

COMPOUND ....

ELECTRON ....

ELEMENT ....

ENERGY SHELL ....

GAS ....

LIQUID ....

MIXTURE ....

MOLECULE ....

NEUTRON ....

NUCLEUS (1) ....

PROTON ....

SOLID ....

WATER ....

Fold

Fold

Workbook 4.2

Draw a Bohr model of each of the atoms listed below. See Figure 4.3 in the text for a sample Bohr model. Refer to Figure 4.4 in the text for the number of subatomic particles in each kind of atom. Remember, only two electrons can fit into the first energy shell, and only eight electrons can fit into the second energy shell.

1. Hydrogen

2. Chlorine

3. Nitrogen

4. Oxygen

Chapter 4

Workbook 4.3

Fill in the blanks with the correct word.

1. Neutrons have .... charge and a mass of one amu (amu = atomic mass unit).

2. Protons have a .... charge and a mass of one amu.

3. Electrons have a .... charge and almost no mass.

4. .... is the most abundant element in your body.

5. The air is 78% nitrogen, an element needed in your body to make ....

6. .... is an explosive gas when in its diatomic form.

7. Organic compounds contain carbon, which is usually a .... color as a solid.

8. The element .... is needed for combustion.

9. Letters are to words as atoms are to ....

10. .... are good conductors of electricity and heat.

11. Noble gases, the most stable elements, are found in Group .... of the Periodic Table.

CHECK IT OUT!   Periodic Table of Videos: *http://www.periodicvideos.com/*

Workbook 4.4

Use the graph below to make a bar graph showing the percent of each element in the human body. Refer to Figure 4.4 in the text for your data.

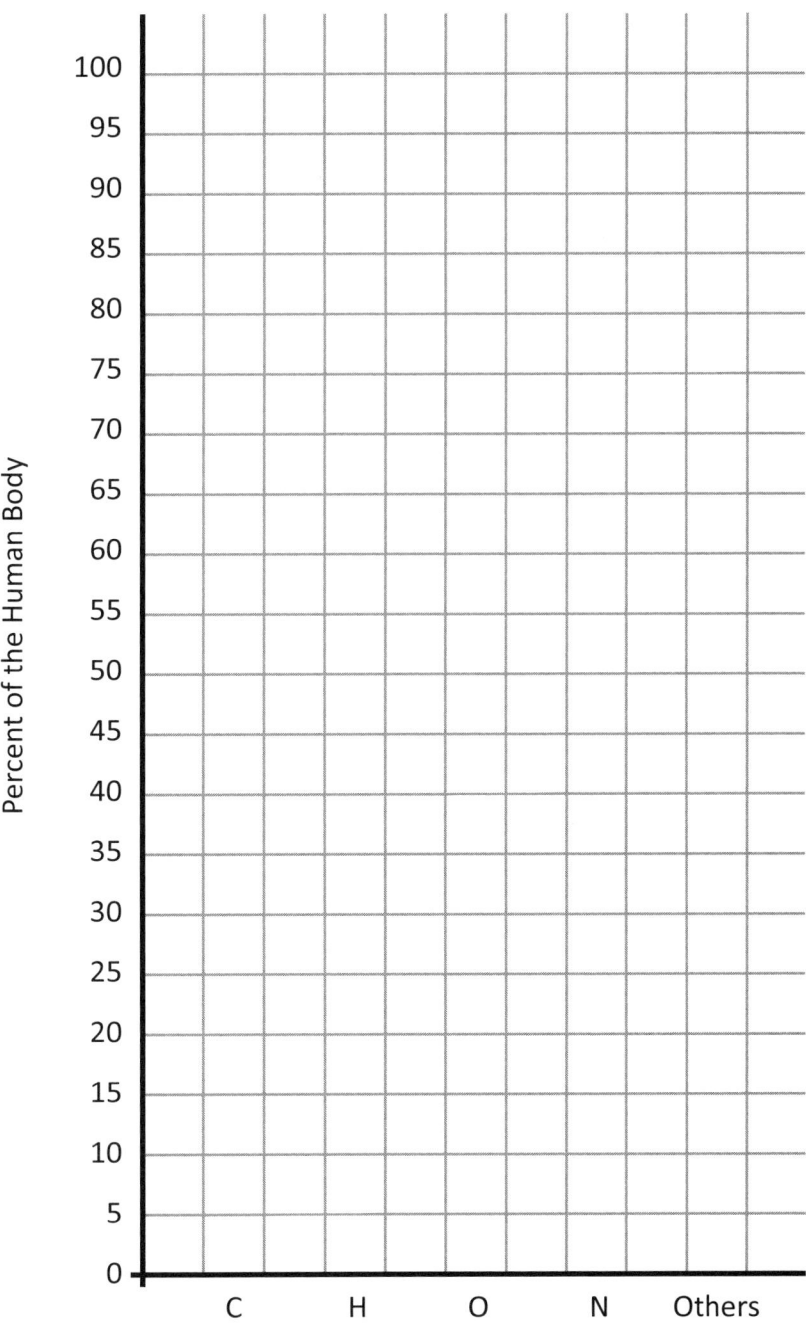

Workbook 4.5

Answer the questions below using complete sentences.

1. List the following seven terms in correct order from smallest to largest:

    molecule   atom   human   neutron   virus   nucleus   electron

    ....

    ....

2. Which part of an atom determines what kind of an atom it is?

    ....

    ....

3. What is most of an atom's volume?

    ....

    ....

4. What is the formula for water?

    ....

    ....

5. How many atoms are in one molecule of water?

    ....

    ....

6. How many elements are in pure water?

    ....

    ....

7. Give three examples of elements.

    ....

    ....

8. List three examples of compounds.

    ....

    ....

9. Name three mixtures.

   ....

   ....

10. What is the one mixture on earth that is not in living things but is most like them?

    ....

    ....

Workbook 4.6

Fill in the blanks with the full name of each element.

1. Ca ....
2. C ....
3. Cl ....
4. Cu ....
5. H ....
6. I ....
7. Fe ....
8. Mg ....

9. N ....
10. O ....
11. P ....
12. K ....
13. Na ....
14. S ....
15. Zn ....

Workbook 4.7

Microscope: Salt Crystals

Supplies:
- microscope
- microscope slide and cover slip
- table salt
- small drinking glass
- eyedropper
- water

1. Make a saturated salt water solution by stirring salt into ¼ cup of water in a small glass. Keep adding salt until no more will dissolve.

2. Using an eyedropper, put one drop of the solution onto a microscope slide. Gently lower a cover slip onto the slide.

3. View your slide through the microscope, following the instructions on page 25. View different areas of your slide, watching for salt crystals to form. The first crystals will probably form on the edge of the cover slip. Be patient! It may take a few minutes for crystals to begin to form.

4. In the space below, draw the salt crystals as you see them through the microscope. Be sure to label your drawing and record the magnification of the microscope. (See page 24 for instructions for determining the magnification of your microscope.)

Title: ....

Magnification: ....

Notes about specimen or procedure:

....
....
....
....
....
....
....
....

Chapter 5

Workbook 5.1

Use the glossary to define each of the keywords on the lines below. Quiz yourself on the keywords every day until you can explain the meaning of each term without looking at the definition.

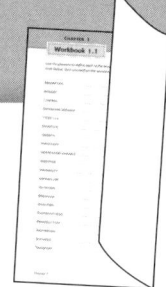

Sample

CELL ....

CELL MEMBRANE ....

CELL WALL ....

CENTRIOLES ....

CHLOROPHYLL ....

CHLOROPLAST ....

CHROMOSOMES ....

CYCLOSIS ....

CYTOPLASM ....

DIFFUSION ....

ENDOPLASMIC RETICULUM ....

GOLGI BODIES ....

LYSOSOME ....

MICROORGANISMS ....

MITOCHONDRION ....

NUCLEAR MEMBRANE ....

NUCLEOPLASM ....

NUCLEUS (2) ....

OSMOSIS ....

PROTOPLASM ....

RIBOSOMES ....

SEMIPERMEABLE ....

VACUOLES ....

Workbook 5.2

Label each term with:

N if it is part of the nucleus,

C if it is a part outside the nucleus,

P if it is a process carried out by the cell.

1. .... diffusion
2. .... chromosome
3. .... ribosome
4. .... mitochondrion
5. .... lysosome
6. .... nucleoplasm
7. .... golgi bodies
8. .... cytoplasm
9. .... digestion
10. .... cyclosis
11. .... chloroplast
12. .... osmosis
13. .... nuclear membrane
14. .... vacuole
15. .... photosynthesis
16. .... cell wall
17. .... endoplasmic reticulum
18. .... protein synthesis
19. .... nucleolus
20. .... cell membrane
21. .... centriole

CHECK IT OUT!   http://www.johnkyrk.com/CellIndex.html
http://www.johnkyrk.com/DNAanatomy.html

Workbook 5.3

Answer the following questions using complete sentences.

1. Which parts are found in plant cells and not animal cells?

   ....

   ....

2. Which part is found in animal cells and not plant cells?

   ....

   ....

3. Which part is usually large in plant cells but small in animal cells?

   ....

   ....

4. Which part holds the instructions for operating the cell?

   ....

   ....

5. Which part is also called the cell membrane?

   ....

   ....

6. Which part is colored green?

   ....

   ....

7. Which part is needed in photosynthesis?

   ....

   ....

8. Which part is the site where proteins are synthesized, or made?

   ....

   ....

9. Which part produces and releases energy to power the cell?

   ....

   ....

10. Which part is the area where food molecules are broken down?

    ....

    ....

11. Which part is for storing proteins?

    ....

    ....

12. Which part controls entry to the nucleus?

    ....

    ....

13. Which process is a circular movement?

    ....

    ....

14. Which process is the movement of water through a membrane?

    ....

    ....

Workbook 5.4

Answer the following questions using complete sentences.

1. Who looked at dead cork through a microscope and used the term "cells" to describe what he saw?

   ....
   ....

2. What is the literal meaning of the word "cell"?

   ....
   ....

3. What is diffusion?

   ....
   ....
   ....

4. Is energy expended in diffusion?

   ....
   ....

5. Is diffusion passive or active transport?

   ....
   ....

6. What is osmosis?

   ....
   ....

7. List two statements of the cell theory.

   ....
   ....
   ....

8. How many cells was your body when your life began?

   ....
   ....

Workbook 5.5

For each cell part pictured below, fill in the blanks with its name (left column) and/or function (right column). Refer to Figures 5.6 and 5.8 in the text.

NAMES AND FUNCTIONS OF CELL ORGANELLES

In the nucleus:

.... 1

Nucleoplasm 2

Chromosomes 3

Outside the nucleus:

.... 4

.... 5

.... 6

.... 7

Ribosomes 8

.... 9

Lysosome 10

Only in animals:

.... 11

In plants only:

.... 12

.... 13

FUNCTION

1. To hold the nucleus together

2. Medium for activity in the nucleus

3. ....

4. Medium for activity in the cell

5. ....

6. ....

7. Surface for chemical activity

8. ....

9. ....

10. ....

11. Helps cells make copies of themselves

12. ....

13. ....

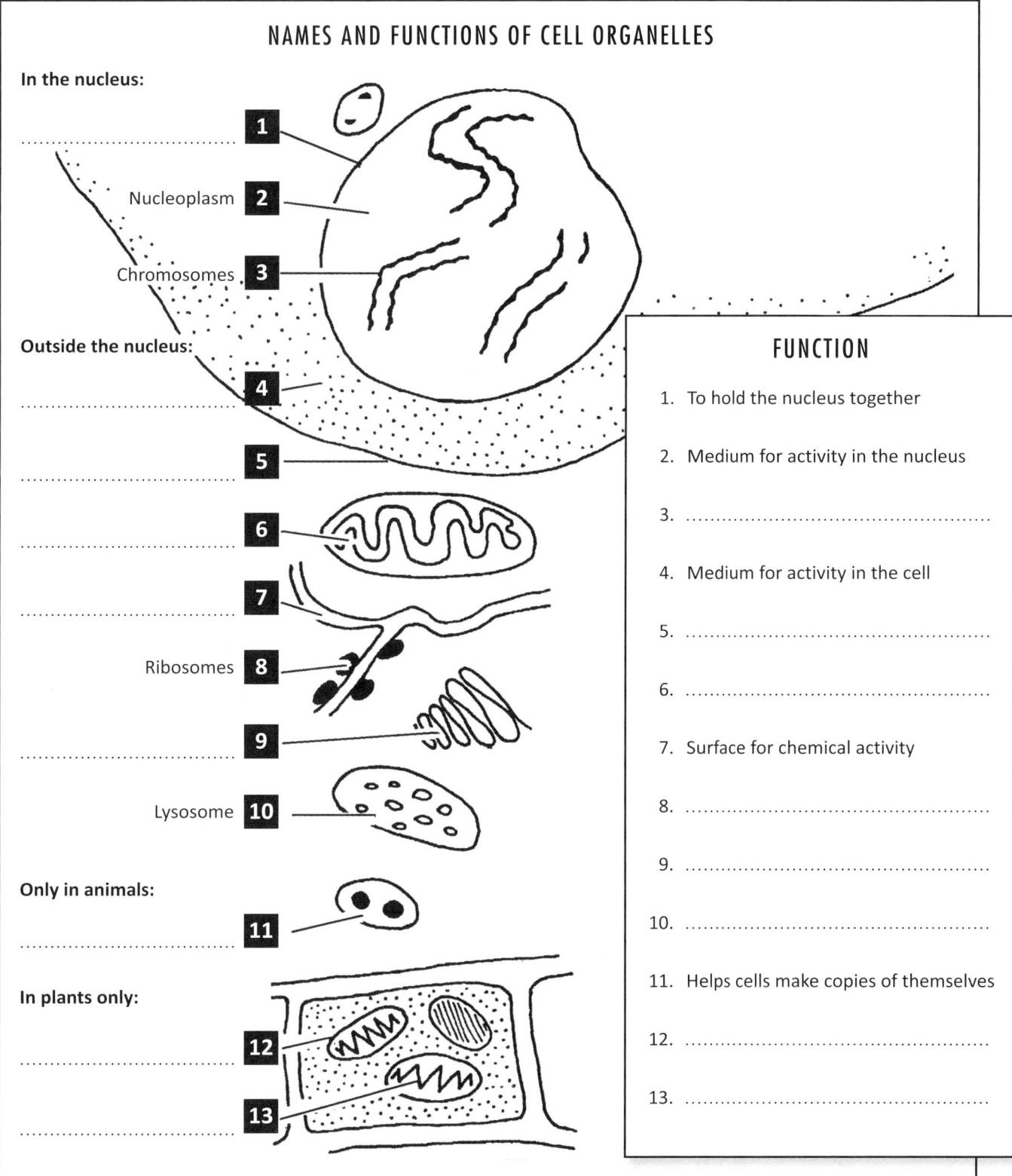

Chapter 5

45

Workbook 5.6

Learning about Osmosis

Supplies:
- drinking glass or mug
- raw egg in shell
- vinegar
- water
- spoon

1. Pour vinegar into the drinking glass until it is almost one inch deep. Place the egg in the bottom of the glass, propping it on one end with a spoon. Approximately ½ of the egg should be immersed in the vinegar.

2. Let the egg sit in the vinegar for about 10 hours or overnight. When the immersed part of the shell is dissolved, remove the egg from the vinegar and carefully rinse the egg.

3. Empty the glass of vinegar and fill it with water. Completely submerge the egg in the water. Osmosis will be evident as the exposed membrane of the egg stretches outward to allow water to move into the egg.

4. Using what you have learned about osmosis, briefly explain the results of your experiment.

   ....
   ....
   ....
   ....
   ....

Workbook 5.7

Microscope: Onion Cells

Supplies:
- microscope
- microscope slide and cover slip
- piece of raw onion (white or yellow)
- eyedropper
- water
- tincture of iodine or Lugol's solution (optional)

Title: ....

Magnification: ....

Notes about specimen or procedure:

....
....
....
....
....
....
....
....
....

1. Take a small piece of onion and peel off the thin membrane from the underside of one of the onion layers.

2. Place the onion membrane flat on the surface of a clean microscope slide. Use an eyedropper to add one drop of water.

3. Gently lower a cover slip onto the slide. To avoid creating air bubbles between the slide and the cover slip, lower the cover slip onto the slide at an angle, allowing one edge to touch the slide first.

4. Follow the instructions on page 25 to view the prepared slide through the microscope.

5. Repeat steps 1–4, this time preparing the slide with a drop of tincture of iodine or Lugol's solution instead of a drop of water. The iodine will stain the vacuoles in the onion cells, making them easier to see.

6. Draw the onion cells in the space below. Be sure to label your drawing and record the magnification of the microscope. (See page 24 for instructions for determining the magnification of your microscope.)

5.8 Formal Lab #2

Complete the experiment below, following the instructions provided. Fill in any blanks as you come to them. Use complete sentences to answer the questions at the end.

Name: .... **Date:** ....

I. Title: Osmosis

II. Purpose: The cell membrane encloses the cytoplasm of every cell and controls entry and exit from the cell. Not all substances can pass through the cell membrane, which is why it is called semipermeable ("Semi" means "halfway" or "partially."). Osmosis is a form of passive transport in which water passes through the cell membrane. By allowing water to pass in and out of the cell, the cell membrane controls the concentration of dissolved chemicals in the cytoplasm.

The cytoplasm of a cell is mostly water, but it also contains dissolved chemicals necessary for life. When a cell absorbs water, it becomes firmer and more rigid, just like a full water balloon. When water leaves a cell, the cell becomes less firm, which is why plants wilt when they don't have enough water. We can use the relative firmness of a potato to determine whether it has absorbed or released water through the process of osmosis.

Before formulating your hypothesis, reread the section on osmosis in Chapter 5 of your textbook.

My hypothesis is that when potato cells are exposed to a saturated salt water solution, water will .... **a) enter the cell, b) exit the cell, or c) neither enter nor leave the cell.**

When potato cells are exposed to pure water, water will .... **a) enter the cell, b) exit the cell, or c) neither enter nor leave the cell.**

III. Materials: The materials required for this experiment are two potatoes, approximately ¼ cup salt, water, and plastic wrap.

IV. Apparatus: The equipment required for this experiment is a paring knife, a cutting board, two small bowls, a 1-cup measuring cup, and a spoon.

V. Procedure:

1. Fill the cereal bowls with one cup of water each. Make a saturated salt water solution in one of the bowls by stirring in as much salt as will dissolve. Keep adding salt, one spoonful at a time, until no more salt will dissolve.

2. Scrub the potatoes at the sink to rinse off the dirt. On a cutting board, use the paring knife to cut each potato in half lengthwise. Wrap one half tightly in plastic wrap and set it aside. This is the control. The plastic wrap will ensure that no water is lost from the potato through evaporation.

3. Place another half in the bowl of salt water and another half in the bowl of pure water. The potato does not need to be submerged entirely, but the cut side should be under water.

4. Set the bowls aside. After about three hours, remove the potatoes from the bowls and remove the plastic wrap from the third potato half.

5. Examine the firmness of the three pieces of potato. Is one of the halves easier or harder to bend? Do any appear to have shrunk or expanded? To test if any have expanded, try fitting two of the halves back together. Use your observations to fill out Figure 1 by describing the firmness and texture of each potato half.

6. When you have completed the experiment, you may wish to use the four potato halves to make mashed potatoes or another delicious dish for dinner!

VI. Data:

	Description
Potato soaked in pure water	
Potato soaked in salt water	
Control	

FIGURE 1. FIRMNESS OF POTATO HALVES

VII. Questions:

1. Did osmosis cause water to enter or leave the cells of the potato that was soaked in pure water? Why?

....

....

....

....

Chapter 5

2. Did osmosis cause water to enter or leave the cells of the potato that was soaked in salt water? Why?

   ....

   ....

   ....

   ....

   ....

3. The **control** in this experiment is the potato wrapped in plastic wrap. This allows us to compare the other potatoes to a potato which has not been soaked in anything. Why is this important?

   ....

   ....

   ....

   ....

   ....

VIII. **Conclusion:** This experiment showed that when potato cells are exposed to a saturated salt water solution, water will pass through the cell membrane .... [into/out of] the cell. When potato cells are exposed to pure water, water will pass through the cell membrane .... [into/out of] the cell.

Thus, my hypothesis was .... **[correct/incorrect/partly correct].**

Chapter 6

Workbook 6.1

Use the glossary to define each of the keywords on the lines below. Quiz yourself on the keywords every day until you can explain the meaning of each term without looking at the definition.

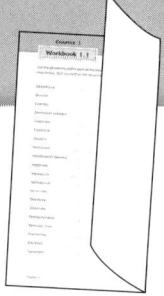

Sample

Aerobic ....

Anaerobic ....

Archaea ....

ATP ....

Autotroph ....

Bacteria ....

Bacterial Spore ....

Bacteriophage ....

Binary Fission ....

Blue-Green Algae ....

Colony ....

DNA ....

Eukaryote ....

Flagellum ....

Heterotroph ....

Humus ....

Mutualism ....

Nodules ....

Parasitism ....

Pasteurization ....

Prokaryote ....

Saprophytism ....

Virus ....

Workbook 6.2

Using a full sentence, list the three basic necessities of life.

....
....
....
....
....

Workbook 6.3

Using complete sentences, list the eight life functions (refer to Chapter 1 if necessary).

....
....
....
....
....
....
....
....
....

Workbook 6.4

Answer each question using a complete sentence.

1. Which is the only life function done by viruses?

2. Which life functions cannot be done by a car?

3. A flagellum is used for which life function?

4. Which life function is the most outstanding characteristic of bacteria?

5. Binary fission is a type of which life function?

6. Where do organisms in Kingdom Archaea thrive?

7. Volvox sp. is a microscopic organism with a cell wall, chloroplasts, and mitochondria. Is Volvox sp. a prokaryote or a eukaryote?

Workbook 6.5

The five environmental conditions required for bacteria to thrive are listed below. Fill in the blanks with the environmental condition associated with each statement.

FOOD   TEMPERATURE   MOISTURE   LIGHT   OXYGEN

1. .... A refrigerator helps keep food fresh.

2. .... Do not use a can of peas with a tiny hole in it.

3. .... A box of cereal is still good after a month.

4. .... Jars are placed in boiling water before being used to preserve jelly.

5. .... A carpet becomes less musty if exposed to sunlight.

6. .... Glucose is an example of this environmental condition.

7. .... Pioneers ate salt pork.

8. .... Potato chips are sold in airtight bags.

9. .... Milk is pasteurized.

10. .... Leaves, birds, algae, and you can provide this environmental condition for bacteria.

Workbook 6.6

Use the keywords from this chapter to fill in the blanks.

1. Chromosomes are made of a double-stranded chemical called ....

2. A .... carries out only one life function: reproduction.

3. Every organism must get energy, which it stores in the cell as .... molecules.

4. The tetanus bacterium is ...., unable to use oxygen gas from the air.

5. .... are organisms without a nuclear membrane.

6. .... is an autotroph which has chlorophyll but no chloroplasts.

7. A bacterium may form a .... to survive drying out.

8. Milk from the farm is sent to the dairy for .... before going to the market.

9. The relationship between nitrogen-fixing bacteria and clover is an example of .... because both the bacteria and the clover benefit from the relationship.

10. A bacterial cell can undergo .... approximately every 30 minutes.

11. Some bacilli use a .... to move around.

12. Without ...., soil is not as fertile.

13. In ...., the microbe harms its host.

14. A .... is formed of so many bacteria that it can be seen without a microscope.

15. Saprotrophic and parasitic bacteria are types of ....

Workbook 6.7

Compose a paragraph explaining four ways bacteria are useful and beneficial for mankind.

6.8 Formal Lab #3

Complete the experiment below, following the instructions provided. Fill in any blanks as you come to them. Use complete sentences to answer the questions at the end.

Name: .... **Date:** ....

I. Title: Comparing the Bacteria on Different Surfaces

II. Purpose: When scientists study bacteria, they often grow them in sterile petri dishes on growth mediums such as nutrient agar. The growth medium provides the bacteria with the nutrients it needs to carry out its life processes, including reproduction. As the bacterial cell divides and multiplies, the cells accumulate in one place to form a circular patch of bacteria. These patches are called colonies, and each one began as a single cell or small group of cells.

In this experiment, we will explore the variety of bacteria found around the house and attempt to determine which surfaces have the most bacteria. The surfaces we will compare are:

1. ....

2. ....

3. ....

4. ....

(Ideas include the inside of your mouth, a door handle, a toilet bowl, the inside of your pet's mouth, a TV remote control, a toothbrush, a comb, beneath your fingernails, etc.)

My hypothesis is that the petri dish cultured with bacteria from .... **will grow the most bacterial colonies.**

⚠ **Safety Note:** The ordinary household bacteria you will collect in this experiment probably will not be harmful. However, even ordinary bacteria can be hazardous when cultured in large quantities. Ask your parent's permission before beginning this experiment, and observe the following safety precautions:

- Do not perform this experiment if any member of your family is sick with a bacterial disease. You don't want to collect the disease-causing bacteria accidentally and culture large quantities of them! (Note: The common cold is a viral disease, not a bacterial disease, so it is all right to perform the experiment if one of your family members has a cold.)

- Once you add the bacteria to the disposable petri dishes, tape the covers onto the dishes so that no one will open them accidentally. After you complete this experiment, dispose of the petri dishes by placing each one, still sealed, inside a ziplock plastic bag. With the assistance of a parent, add ¼ cup bleach and ¼ cup water to each bag, then seal the bag and put it in the trash.

- When you open the petri dishes to wipe off the moisture (step 8), be careful not to ingest, breathe, or touch the bacteria. Do not handle the opened petri dish if you have an open cut on your hand.

- Wash your hands with soap and warm water after handling the petri dishes.

III. **Materials:** The materials required for this experiment are a 125-ml bottle of liquid nutrient agar, five disposable petri dishes, tap water, paper towels, and household bleach.

IV. **Apparatus:** The equipment required for this experiment is a permanent marker, q-tips, tape, and a large ziplock plastic bag.

V. **Procedure:**

1. Pick four surfaces which you would like to test for bacteria. List the surfaces on the lines provided in Part II, and record your hypothesis by filling in the blank.

2. Follow the directions on the bottle of agar to fill five petri dishes with nutrient agar growth medium. Keep the petri dishes closed while you prepare the agar, and only open them when filling each one with agar. Put the cover back on the petri dish as soon as it has been filled.

3. When the agar has solidified, wash your hands with soap and warm water. Moisten a clean q-tip with water and rub it over the first surface on your list. Remove the lid from one petri dish and roll the q-tip over the surface of the agar in the shape of a large "Z." Immediately replace the cover of the petri dish and tape it closed. Turn the petri dish upside down and use the permanent marker to label the bottom of the dish.

4. Repeat step 3 in exactly the same way for the rest of the surfaces on your list. Use a fresh q-tip for each surface and wash your hands before and after inoculating (i.e., adding bacteria to) each petri dish. Don't forget to label each petri dish.

5. The fifth petri dish is the control. After washing your hands, moisten a clean q-tip with water and roll it over the surface of the agar in the shape of a large "Z." *Do not touch the fifth q-tip to any surface before rolling it over the agar*. Tape the petri dish closed and use the permanent marker to write "Control" on the bottom of the dish.

6. Store the dishes upside down in a warm, dark place where they will not be disturbed. (Storing the dishes upside down ensures that any moisture that collects on the covers will not drip down onto the bacteria.)

7. Leave the petri dishes alone for approximately 5–7 days, or until a significant number of bacterial colonies have grown. Try not to turn the dishes right side up during this time.

8. After one week, turn the dishes right side up. Cut the tape and remove the cover of each petri dish long enough to wipe the moisture off the inside of the cover with a paper towel. Put the covers back on the petri dishes and tape them closed again. Put the paper towels in the trash right away to avoid contaminating other surfaces.

9. Study and compare the bacteria in the different petri dishes. You should see many colonies of bacteria. Count the colonies in each petri dish and record the data in Figure 1. In some petri

dishes, the colonies may be crowded together so closely that it will be difficult to distinguish one from the other. If you look closely, however, you should be able to count the dot-like colonies. Instead of counting all the colonies, you can count the colonies in a specific area of the petri dish and then use that number to estimate the colonies on the rest of the dish.

10. After recording your data in Figure 1, select the petri dish which you find most interesting and draw and color it in the space provided for Figure 2. Fill in the blank to complete the title of Figure 2.

11. When you have completed the experiment, ask your parent to pour a little bleach into each petri dish. Allow the dishes to sit overnight, and then rinse the bleach in each petri dish down the sink. Dispose of the decontaminated petri dishes in a large, sealed ziplock bag.

VI. Data:

	Surface from which the bacteria were taken	Number of colonies	Number of different types of colonies
Dish 1			
Dish 2			
Dish 3			
Dish 4			
Dish 5	Control		

FIGURE 1. NUMBER OF COLONIES IN EACH PETRI DISH

FIGURE 2. Bacterial colonies in the petri dish with bacteria taken from ....

VII. Questions:

1. Which petri dish contained the most colonies of bacteria? From what surface were the bacteria in that dish collected?

   ....

   ....

2. Since each bacterial colony began as a single cell or small group of cells, the number of colonies in each petri dish allows us to estimate the number of bacteria that were transfered to the dish from the different surfaces that we tested. Given this information, which surface had the most bacteria?

   ....

   ....

3. In this experiment, we prepared each petri dish in exactly the same way in order to control every variable except the independent variable. For instance, we used the same growth medium and the same method of collecting bacteria for all the petri dishes. If we were scientists performing this experiment in a laboratory or hospital, we would need to take even more precautions to ensure that our results were free of error. For instance, we would moisten the q-tip with sterilized water rather than water from the faucet. List at least one other possible source of error in the experiment and explain how you could change the experiment to eliminate that source of error.

   ....

   ....

   ....

4. Did any bacterial colonies grow on the parts of the agar which were not touched by the q-tip? If so, where do you think these bacteria came from?

   ....

   ....

   ....

   ....

5. Did any bacterial colonies grow in the petri dish labeled "Control"? If so, where do you think these bacteria came from?

   ....

   ....

   ....

VIII. Conclusion:

This experiment showed that the petri dish cultured with bacteria from

.... grew the most bacterial colonies.

Thus, my hypothesis was .... [correct/incorrect].

Chapter 7

Workbook 7.1

Use the glossary to define each of the keywords on the lines below. Quiz yourself on the keywords every day until you can explain the meaning of each term without looking at the definition.

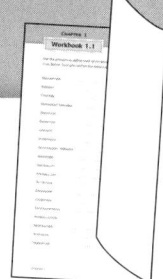

Sample

Binomial Nomenclature ....

Cilia ....

Classes ....

Conjugation ....

Contractile Vacuole ....

Eyespot ....

Families ....

Genera ....

Kingdoms ....

Macronucleus ....

Micronucleus ....

Multicellular ....

Orders ....

Pellicle ....

Phyla ....

Protista ....

Pseudopod ....

Species ....

Taxon ....

Taxonomy ....

Unicellular ....

Fold

Fold

Workbook 7.2

Using your keywords, answer each question using a complete sentence.

1. What is the science of naming creatures?

   ....

2. Which taxon contains just one kind of organism?

   ....

3. What is the term for using two names to identify an organism?

   ....

4. What are the two taxons used in a scientific name?

   ....

   ....

5. Which of the seven taxons is largest?

   ....

6. There are only six of which group?

   ....

7. Families are subdivisions of which group?

   ....

8. Creatures which are almost, but not exactly the same, would be in which groups together?

   ....

   ....

9. Which taxon could contain organisms that are most different?

   ....

10. Different protists in the same phylum must also belong to the same _____ . (Which one of the other taxons?)

    ....

Workbook 7.3

Answer each question using a complete sentence.

1. Which ancient Greek started the first orderly classification of living things?

   ....
   ....

2. What was the most important thing Carolus Linnaeus did for taxonomy?

   ....
   ....

3. Are the house fly and the human in the same kingdom?

   ....
   ....

4. Can the red oak and an unborn baby be in the same species?

   ....
   ....

5. List three rules for writing scientific names correctly.

   ....
   ....

6. What is your scientific name?

   ....
   ....

7. What do scientists use as the basis for classifying organisms?

   ....
   ....

8. Are two organisms in the same genus very similar or very different?

   ....
   ....

Workbook 7.4

Diagram, label, and color a paramecium in the space below. Refer to Figure 7.14 in the text. You may use a pencil to draw the diagram, but use a pen to write the labels. Use a ruler for any straight lines, and don't forget to include a title.

Title: ....

Workbook 7.5

If the statement is true, rewrite it on the lines provided. If the statement is false, change the italicized word to make the statement true. Write the corrected statement on the lines provided.

1. Kingdom *Bacteria* contains unicellular organisms with organelles.

   ....
   ....

2. Slime mold and golden algae belong to the same *phylum*.

   ....
   ....

3. Euglena, amoeba, and paramecium belong to the same *phylum*.

   ....
   ....

4. The amoeba moves by means of a *flagellum*.

   ....
   ....

5. The *paramecium* uses cilia for locomotion.

   ....
   ....

6. The euglena has a *flagellum* to help it perform the life function of irritability.

   ....
   ....

7. The *amoeba* is both plant and animal like.

   ....
   ....

8. The amoeba can reproduce by *binary fission*.

   ....
   ....

9. All *protists* have a nucleus, with a membrane, that controls the cell.

   ....
   ....

10. The oak tree and the paramecium share *two* of the same taxons.

    ....
    ....

11. All the phyla of the Kingdom Protista have organelles, a nuclear membrane, and *flagella*.

    ....
    ....

Workbook 7.6

Microscope: Protists

Supplies:
- microscope
- microscope slide and cover slip
- pond water or seawater
- jar or bucket
- eyedropper

1. Collect several cups of water from a pond, ditch, river, or tide pool. Be sure to include some of the algae and decomposing plants at the bottom of the pond. Allow the water to sit in the open jar or bucket for at least two days. Keep the water at close to room temperature, and avoid placing it in direct sunlight for extended periods of time. During this time, the protists that naturally live in pond water will multiply, making it easier to find them in a drop of water.

2. Gently stir the jar of water, then use the eyedropper to place one small drop of water onto the slide. Cover gently with a cover slip.

3. Follow the instructions on page 25 to view the prepared slide through the microscope. If you have difficulty focusing the microscope, try adjusting the diaphragm. Most of the objects you will see are bits of algae and dead plants. Look carefully, though, and you should see tiny protists darting around the slide! It is easiest to spot them on low power, because you can see more of the slide at once. When the slide begins to dry up, the protists will stop moving, and it will be difficult to find them. You can rinse off the slide and start over with a new drop of pond water.

4. Once you have explored the different types of protists on low power, try viewing one on high power. Start with a new drop of pond water, and cover with a cover slip. When you have spotted a protist on low power, move the slide around the stage to keep the protist in view. As the water on the slide begins to dry up, the protist will slow down. Now you can view the protist on high power. Adjust the diaphragm as needed to get the best view of the tiny creature.

5. Draw one or more of the protists on the next page. Be sure to label your drawing(s) and record the magnification of the microscope. (See page 24 for instructions for determining the magnification of your microscope.)

6. Optional: Repeat steps 1–4 with water from different sources. Is there a difference in the type of protists that live in pond water, in a puddle, in a tide pool, and in river water?

CHECK IT OUT! http://www.microscopy-uk.org.uk/pond/collect.html

Notes about specimen or procedure:

....
....
....
....
....
....
....
....
....
....
....
....

Title: ....

Magnification: ....

Notes about specimen or procedure:

....
....
....
....
....
....
....
....
....
....
....
....
....
....

Title: ....

Magnification: ....

Chapter 8
Workbook 8.1

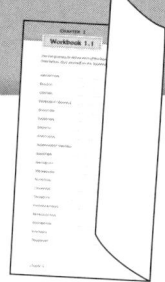
Sample

Use the glossary to define each of the keywords on the lines below. Quiz yourself on the keywords every day until you can explain the meaning of each term without looking at the definition.

Don't forget to review your keywords from Chapters 1-7! Frequent review of keywords from previous chapters is critical to doing well on the Midway Review and Final Review.

ANNULUS ....

ASCUS ....

BASIDIA ....

BUDDING ....

CAP ....

CHITIN ....

FUNGI ....

GILLS (1) ....

HYPHAE ....

MYCELIUM ....

MYCOLOGIST ....

RHIZOIDS ....

SPORE ....

STIPE ....

STOLON ....

VOLVA ....

Workbook 8.2

Using complete sentences, explain how fungi are different from archaeans and bacteria.

....
....
....
....
....
....
....
....

Workbook 8.3

Using complete sentences, explain how fungi are different from protists.

....
....
....
....
....
....
....
....

Workbook 8.4

A. Using a complete sentence, list four characteristics of Kingdom Fungi.

....
....
....
....
....
....
....

B. Compose a paragraph describing five ways that fungi are helpful or harmful to human beings.

....
....
....
....
....
....
....
....
....
....
....

Workbook 8.5

Fill in the blanks with the best answer.

1. The .... are plant-like organisms without true roots, stems, or leaves and no chlorophyll.

2. A .... is fascinated by the variety and weirdness of fungi, and so this person studies them in detail.

3. Fungi must obtain their food energy from other organisms or decaying organic matter because they are ....

4. The ring on some mushrooms is called an .... from the same root word that is in "annual," meaning one ring around the sun.

5. The .... holds the cap up off the ground.

6. Almost all fungi are made of thin strands called ....

7. Blue-green algae have a cellulose cell wall but fungi have .... in their cell walls.

8. The sac-shaped structure containing spores in the ascomycetes is called an ....

9. In the basidiomycetes, the spores form on club-shaped structures called ....

Workbook 8.6

Diagram, label, and color a typical mushroom. Refer to Figure 8.6 in the text. You may use a pencil to draw the diagram, but use a pen to write the labels. Use a ruler for any straight lines, and don't forget to include a title.

Title: ....

Workbook 8.7

Spore Prints

Supplies:
- one **freshly picked, mature** mushroom (its gills must be visible when it is picked)
- one sheet of white paper
- one sheet of black paper

1. Carefully cut or break off the mushroom's stipe. Place the mushroom cap on the paper so that it is half on the white sheet and half on the black sheet. (Depending on the color of the spores, the print may show up better on the black paper or on the white paper.) The mushroom's gills should be facing down.

2. Put the mushroom and paper where they will not be disturbed. The next day, carefully lift the mushroom cap off the paper to observe the spore print.

   ⚠ Caution: Many mushrooms are poisonous. Always wash your hands after handling mushrooms you have found in the wild!

Workbook 8.8

Microscope: Mushroom Spores

Supplies:
- microscope
- microscope slide and cover slip
- one freshly picked, mature mushroom (its gills must be visible when it is picked)
- water
- eyedropper

1. Carefully cut or break off the mushroom's stipe and place the mushroom cap on the microscope slide with its gills facing down. Put the mushroom and the slide where they will not be disturbed.

2. The next day, remove the mushroom and cover the slide with a cover slip, being careful not to smudge the spores. Examine the spores under the microscope, following the instructions on page 25. Remember to begin with the low power objective lens and then proceed to the high power lens. The spores will be visible as round specks or balls.

Workbook 8.9

Microscope: Yeast Cells

Supplies:
- microscope
- microscope slide and cover slip
- active dry yeast
- sugar
- water
- eyedropper
- small bowl

1. Stir 2 ¼ tsp. of yeast and 1 Tbs. of sugar into ⅓ cup of warm water. Make sure the water is warm, but not hot. (Hot water will kill the yeast cells.) When the sugar is dissolved, set the mixture aside until it begins to bubble and foam, about 5–10 minutes.

2. Place one drop of the yeast mixture onto the microscope slide and cover it with a cover slip. Examine the slide under the microscope, following the instructions on page 25. Remember to begin with the low power objective lens and then proceed to the high power lens. You should see hundreds of tiny, round yeast cells floating in the water.

3. Draw what you see at high power in the space below. Be sure to label your drawing and record the magnification of the microscope. (See page 24 for instructions for determining the magnification of your microscope.)

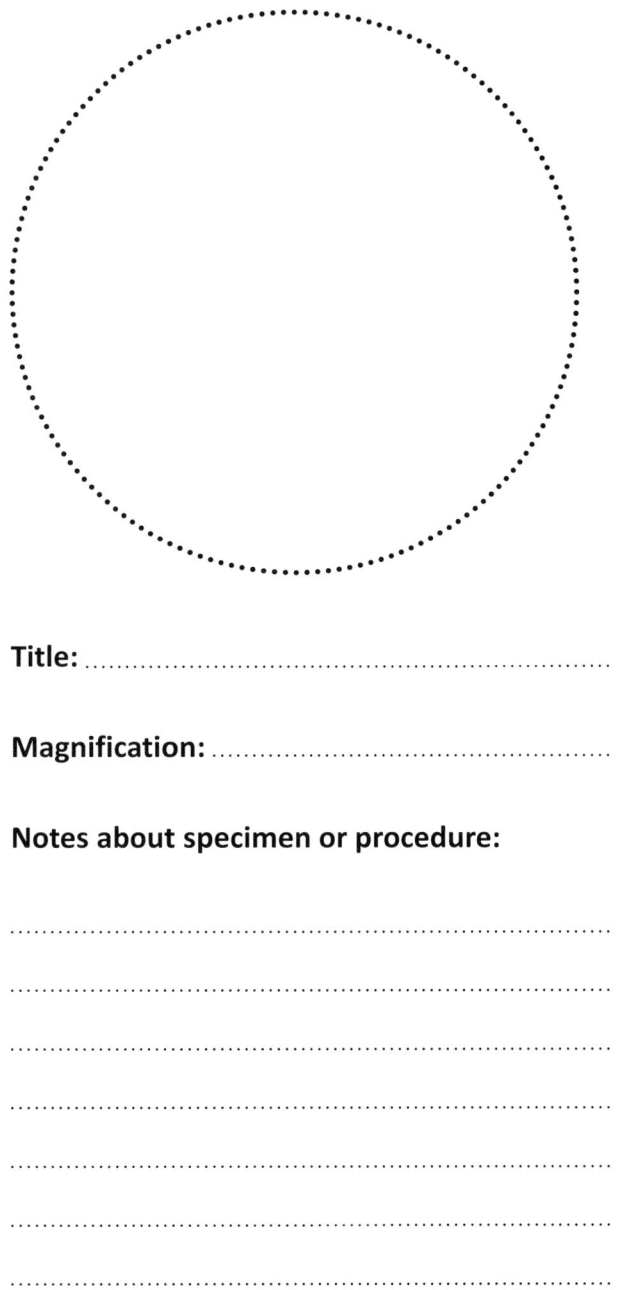

Title: ....

Magnification: ....

Notes about specimen or procedure:

....
....
....
....
....
....
....
....
....

Just for Fun!

NOTICE: CONTINUE READING ONLY IF YOU HAVE A TOADSTOOL TEMPERAMENT, A MUSHROOM MENTALITY, OR A MUSTY, MOLDY, MYCOLOGICAL MADNESS!

The Kingdom Fungi is well known for including important decomposers. There is a fungus among'us, Ann Ulus, who wishes to ascus a few rotten riddles.

1. Why did Average Adam, with his toy gun, need some mushrooms?

2. Why were the police investigating the fan-shaped hyphae?

3. What did the yeast cell say to the baker?

4. Why did Haphazard Harriet think fungi were fish?

5. What do you call a boy who is always joking?

6. What did Average Arnold say when he had to pay a lot to the doctor?

7. Why couldn't the cell in the ascus pay to go to the movies?

8. What did Mrs. Franklin tell Benjamin when it began to thunder?

9. What is the name of the special class where Eskimos teach their dogs to pull their dog sleds?

Answers

1. He needed CAPS.
2. It was STOLON (stolen).
3. This BUD's for you.
4. They have GILLS.
5. FUNGI (fun guy).
6. That was a HYPHAE (high fee).
7. Because she'SPORE (she's poor).
8. Bring your CHITIN (kite in).
9. MUSHROOM ("mushi" room)

Chapter 8

Chapter 9

Workbook 9.1

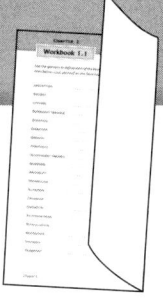
Sample

Use the glossary to define each of the keywords on the lines below. Quiz yourself on the keywords every day until you can explain the meaning of each term without looking at the definition.

ANGIOSPERMAE ....

BRYOPHYTA ....

CONIFER ....

COTYLEDON ....

DICOT ....

FILICINEAE ....

FROND ....

GYMNOSPERMAE ....

LICHEN ....

MONOCOT ....

PLANTAE ....

PYRENOID ....

RHIZOME ....

SORI ....

SPORANGIA ....

TRACHEOPHYTA ....

VASCULAR SYSTEM ....

Workbook 9.2

Using a complete sentence, explain how plants are different from:

1. Prokaryotes:

   ....

   ....

2. Protists:

   ....

   ....

3. Fungi:

   ....

   ....

4. Animals:

   ....

   ....

Workbook 9.3

Diagram, label, and color a fern in the space below. Refer to Figure 9.20 in the text. You may use a pencil to draw the diagram, but use a pen to write the labels. Use a ruler for any straight lines, and don't forget to include a title.

Title: ....

Workbook 9.4

If the statement is true, rewrite it on the lines provided. If the statement is false, change the italicized word to make the statement true. Write the corrected statement on the lines provided.

1. *Heterotroph* means "self feeder."

   ....
   ....

2. In photosynthesis, plants use light energy to make *food*.

   ....
   ....

3. Plants have nuclear membranes, *fungi* do not.

   ....
   ....

4. Plants have *chitinous* cell walls, fungi do not.

   ....
   ....

5. Plants have *chlorophyll* but fungi do not.

   ....
   ....

6. Phylum Chlorophyta is the *green* algae.

   ....
   ....

7. *Chlorella sp.* has a spiral chloroplast.

   ....
   ....

8. *Sugar,* in the pyrenoids, turns iodine blue-black.

   ....
   ....

9. Lichens, containing both algae and fungi, are a good example of *transpiration.*

   ....
   ....

10. *Lichens* are organisms that can live on bare rocks.

    ....
    ....

Workbook 9.5

Compose a short paragraph explaining why ferns usually grow only where it is shady and moist. Include an explanation of why ferns can grow taller than mosses.

....
....
....
....
....
....
....
....
....

Workbook 9.6

Fill in the blanks with the best answer.

1. Liverworts, hornworts and mosses are in Phylum ....

2. Mosses tend to be short because they have no ....

3. All bryophytes need a film of water for ....

4. What appear to be tiny roots on a moss are actually ....

5. Haircap moss is in Kingdom ....

6. Haircap moss is in Phylum ....

7. Haircap moss is in Class ....

8. The genus for haircap moss is ....

9. The species name for the haircap moss is ....

10. .... is the scientific name for the haircap moss.

Workbook 9.7

Fill in the blanks with the best answer.

1. Ferns belong to the Kingdom ....

2. The New York fern is a member of the Phylum ....

3. Unlike mosses, ferns have a .... for transport of liquids.

4. The fan-like leaf of a fern is called a ....

5. A horizontal underground stem that connects some ferns is a ....

6. .... are small dots on the undersides of certain pinnae.

7. The .... contain a number of spores, and are clustered in the sori.

8. The common bracken fern is in the Class ....

9. Instead of seeds, ferns reproduce by ....

10. Because ferns make their own food we can call them ....

Workbook 9.8

Answer each question with a complete sentence.

1. What is the one characteristic all tracheophytes have in common?

   ....

   ....

2. Which of these three main classes of Phylum Tracheophyta do not produce seeds: Filicineae, Gymnospermae and/or Angiospermae?

   ....

   ....

3. What is the meaning of the root word "*gymno?*"

   ....

   ....

4. What special plant organ do plants in Class Angiospermae have that plants in Class Gymnospermae do not have?

   ....

   ....

5. What class contains the conifers?

   ....

   ....

6. What class is also called the flowering plants?

   ....

   ....

7. What are the seeds hidden inside of, in the flowering plants?

   ....

   ....

8. Give two important uses of conifers.

   ....

   ....

9. What is the meaning of the subclass name Monocotyledonae?

   ....

   ....

10. Are dicots angiosperms?

    ....

    ....

Workbook 9.9

Microscope: Green Algae

Supplies:
- microscope
- microscope slide and cover slip
- green algae
- transparent glass or jar
- tweezers
- eyedropper
- water
- Lugol's solution or tincture of iodine

1. Collect some green algae from a pond or other location. If you collected it from a shady place, put the algae in water in a transparent glass or jar and set it in the sun for several hours to make sure the pyrenoids are full of starch from the process of photosynthesis.

2. Use tweezers to place a few filaments of algae on a microscope slide, and add one drop of water with the eyedropper. Place a cover slip over the algae.

3. Follow the instructions on page 25 to view the prepared slide through the microscope. The algae will look like green strings across the slide.

4. Switch to high power and focus on one of the "strings" of algae. You should be able to identify the cell walls and green chloroplasts of the algae cells. You may also be able to identify the nuclei and vacuoles in the cells.

5. Prepare another slide with new filaments of algae. This time, add a tiny drop of tincture of iodine or Lugol's solution to the slide instead of water. The iodine will stain the starch that is stored in pyrenoids on the chloroplasts.

6. Follow the instructions on page 25 to view the prepared slide through the microscope, beginning on low power and proceeding to high power. Can you see the iodine-stained pyrenoids on the chloroplasts?

7. Draw what you see at high power in the space below. Be sure to label your drawing and record the magnification of the microscope.

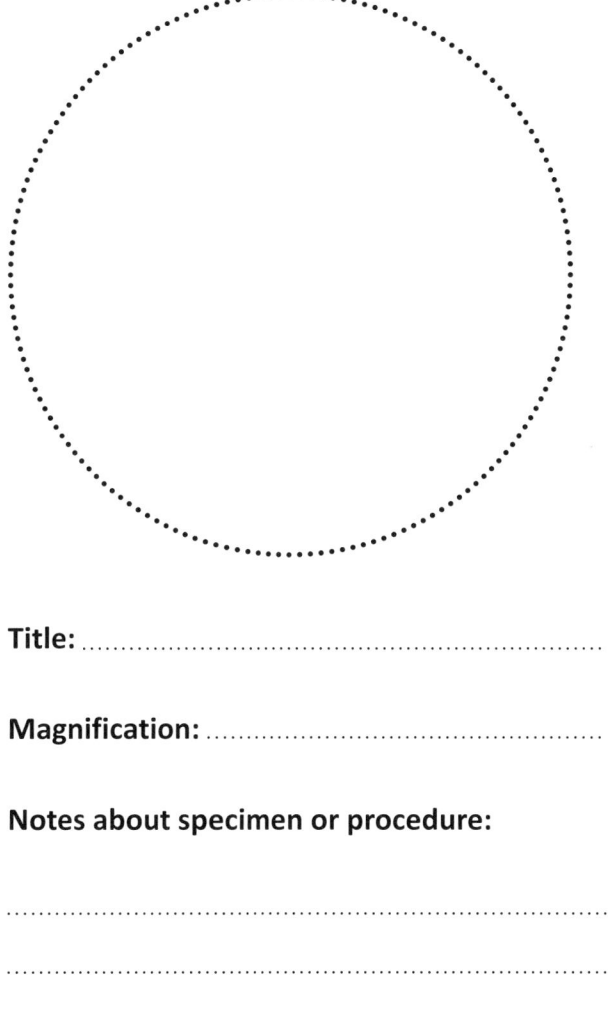

Title: ....

Magnification: ....

Notes about specimen or procedure:

....

....

....

....

Workbook 9.10

Diagram, label, and color a plant cell in the space below. Refer to Figure 5.6 in the text. You may use a pencil to draw the diagram, but use a pen to write the labels. Use a ruler for any straight lines, and don't forget to write a title.

Title: ....

Chapter 10

Workbook 10.1

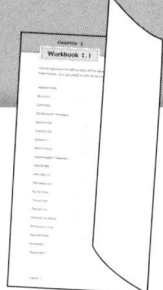

Sample

Use the glossary to define each of the keywords on the lines below. Quiz yourself on the keywords every day until you can explain the meaning of each term without looking at the definition.

AUXINS ....

BARK ....

BULB ....

CAMBIUM ....

CUTICLE ....

FIBROUS ROOT ....

GEOTROPISM ....

HERBACEOUS STEMS ....

LENTICELS ....

PALISADE LAYER ....

PHLOEM ....

PHOTOTROPISM ....

PITH ....

ROOT CAP ....

ROOT HAIRS ....

SPONGY LAYER ....

STOMATE ....

TAP ROOT ....

TRANSPIRATION ....

TROPISM ....

TUBER ....

WOODY STEMS ....

XYLEM ....

Workbook 10.2

Neatly rewrite the following terms in order from smallest to largest:

atoms   cells   molecules   organelles   organism   organs   systems   tissues

1. ....
2. ....
3. ....
4. ....
5. ....
6. ....
7. ....
8. ....

Workbook 10.3

Compose a short paragraph that explains where guard cells may be found, what their function is, and how their shape helps them do their function. Include a title and correctly use five of these terms in your paragraph: flaccid, guard cell, lower epidermis, spongy layer, stomate, transpiration, turgid.

....
....
....
....
....
....
....
....
....

Workbook 10.4

A. *Fill in the blanks with the best answer.*

1. Many similarly shaped cells form a ...., which in turn forms an organ.

2. A .... system is very good for absorption.

3. A carrot is an example of a .... in which food is stored.

4. Food flows in the ....

5. Healthy roots have lots of tiny .... to aid in absorption.

6. Plants that live only 1 or 2 years have ....

7. Many perennials have .... on their stems, a thick layer of dead cells for protection of the tissues underneath.

8. The onion is a special underground stem called a ....

9. Most photosynthesis occurs in the ...., made of tall cells under the upper epidermis.

10. Oxygen gets into a leaf through the ....

11. .... is the evaporation of water from plants.

12. The .... and phloem tissues together make up the vascular system of tracheophytes.

13. Plants have .... which cause different cells to grow different amounts.

14. A small tree grows out from under a dark porch because of ....

B. *List five different plant tropisms and the meaning of the root word in each term.*

1. ....

2. ....

3. ....

4. ....

5. ....

Workbook 10.5

Diagram, label, and color a root longisection in the space below. Refer to Figure 10.6 in the text. You may use a pencil to draw the diagram, but use a pen to write the labels. Use a ruler for any straight lines, and don't forget to write a title.

Title: ....

Workbook 10.6

Fill in the blanks to complete the classification chart. Refer to Figures 9.3 and 9.26 in the text.

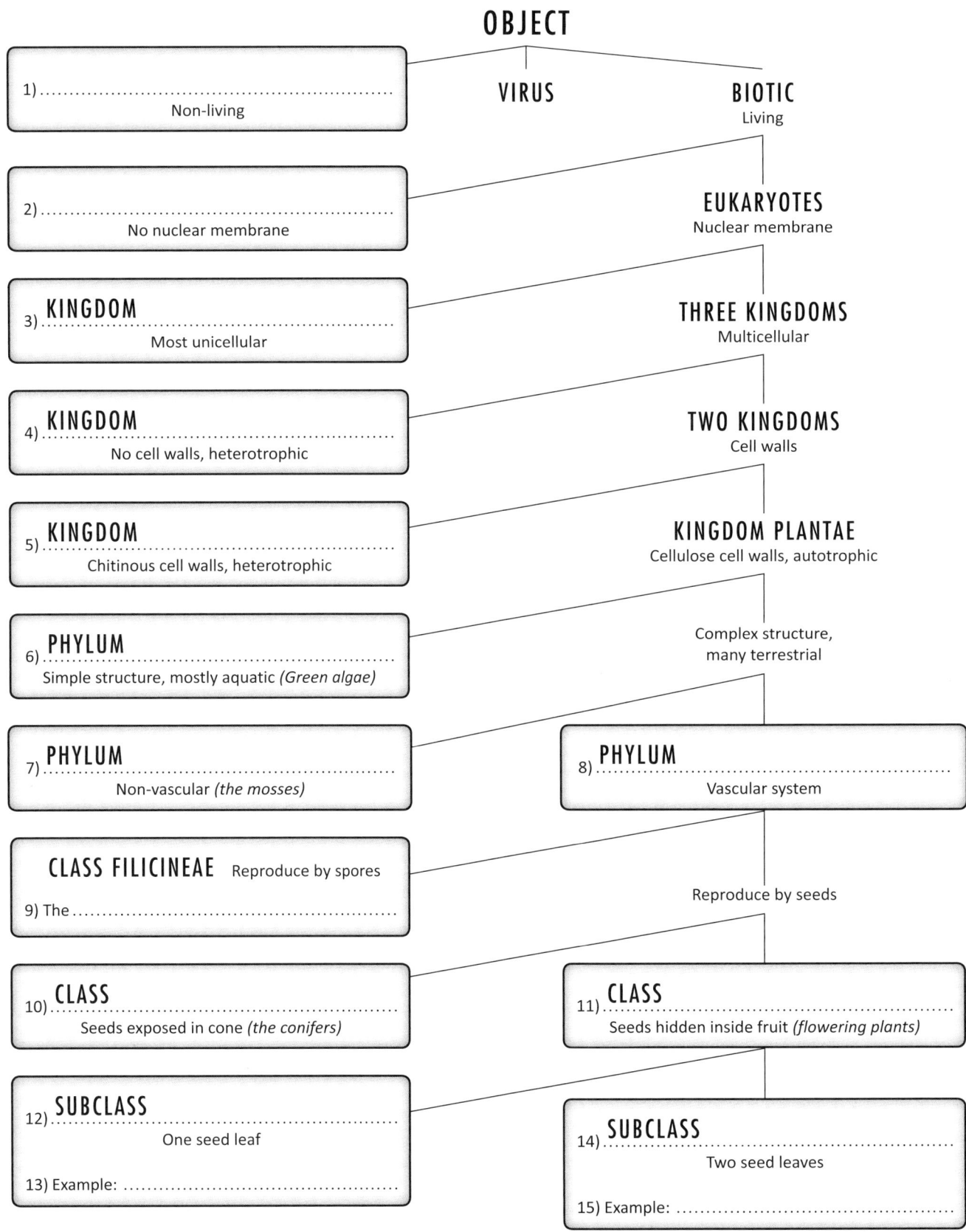

Workbook 10.7

Diagram, label, and color a cross-section of a leaf in the space below. Refer to Figure 10.14 in the text. You may use a pencil to draw the diagram, but use a pen to write the labels. Use a ruler for any straight lines, and don't forget to write a title.

Title: ....

Workbook 10.8

Microscope: Xylem and Phloem in a Celery Stem

Supplies:
- stalk of celery
- paring knife
- tall drinking glass
- water
- food coloring (blue or red)
- spoon
- microscope
- microscope slide and cover slip
- eyedropper

1. Trim one inch off the end of the celery stalk.

2. Fill the glass two-thirds full of water. Add about 10 drops of food coloring and stir.

3. Place the celery stalk in the drinking glass and let it sit overnight, or until the dye has reached the leaves of the celery.

4. Rinse the dye off the celery stalk and cut off its top. Cut an extremely thin sliver off the top of the celery.

5. Place the thin slice of celery on a microscope slide. Put a drop of water on the celery slice and cover with a cover slip. Examine the slide following the instructions on page 25.

6. Can you identify the vascular bundles in the celery stem? The vascular bundles in celery are the "strings" which get stuck in your teeth when you eat raw celery. Since the water and dye are transported up the stem by the xylem, the vascular bundles absorb more dye than the other parts of the celery.

7. Draw what you see at high power in the space below. Be sure to label your drawing and record the magnification of the microscope.

Title: ....

Magnification: ....

Notes about specimen or procedure:

....
....
....
....
....
....
....
....

Workbook 10.9

Microscope: Parts of a Plant

Supplies:
- microscope
- microscope slide and cover slip
- flower petal
- leaf
- root hair
- water
- eyedropper

1. Place the flower petal on a slide and cover with a cover slip. If the whole petal is too thick or large to lie flat under the cover slip, tear off a narrow strip of the petal to use instead. Examine the slide under the microscope, following the instructions on page 25.

2. Tear off a small portion of the leaf and place it upside down on a slide. Put a drop of water on the leaf and cover with a cover slip. Examine the slide following the instructions on page 25.

3. Place the thinnest root hair you can find on a slide. Put a drop of water on it, cover it with a cover slip, and examine the slide under the microscope, following the instructions on page 25.

4. If you wish, you may also view other parts of plants under the microscope. Ideas include a blade of grass, a thin piece of bark, young and mature leaves of the same plant, the leaves of different plants, and petals of different colors.

5. Draw and color your favorite specimen at high power in the space below. Be sure to label your drawing and record the magnification of the microscope. (See page 24 for instructions for determining the magnification of your microscope.)

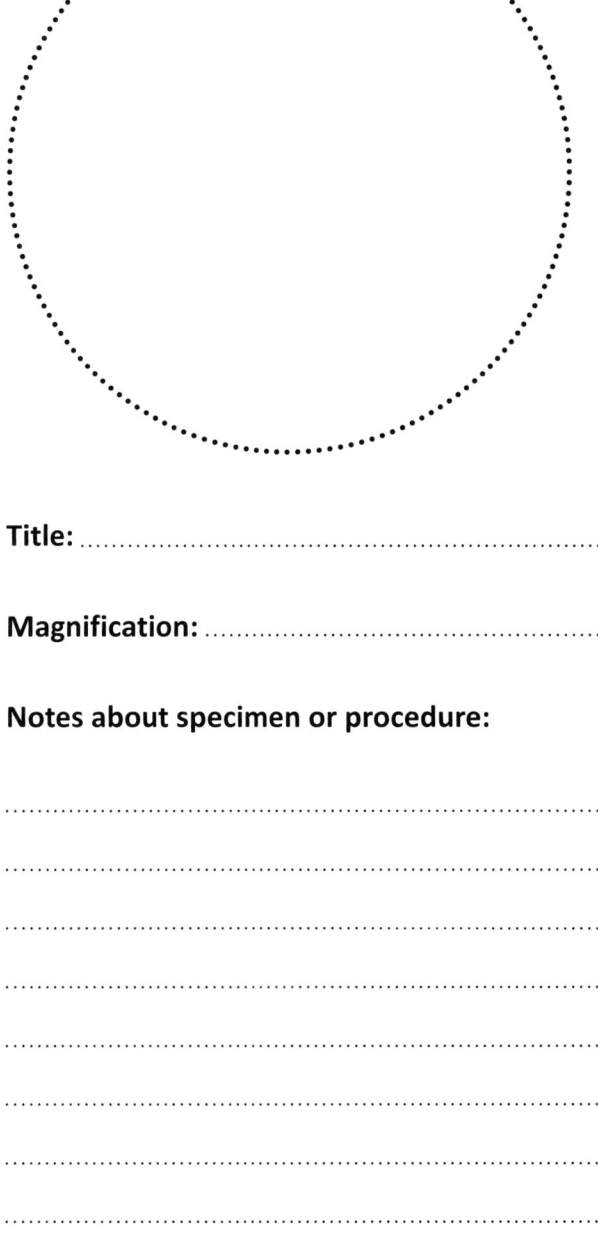

Title: ....

Magnification: ....

Notes about specimen or procedure:

....
....
....
....
....
....
....
....
....

Workbook 10.10

Microscope: Leaf Stomata

Supplies:

- clear nail lacquer (nail *polish* or *enamel* is too thin, and tends to stick to the leaf)
- a sturdy leaf that is not covered with "hairs" (leaves from a rose bush work well)
- microscope
- microscope slide and cover slip
- eyedropper
- water

1. Wash the leaf and dry it well.

2. Outside or in a well-ventilated room, paint the bottom of the leaf with one coat of nail lacquer. You do not have to coat the entire leaf—an area about 1-cm square will be sufficient. Allow the lacquer to dry completely.

3. When the lacquer is dry, peel it off the leaf. The lacquer will retain a textured impression of the leaf.

4. Place the dry nail lacquer on a microscope slide. Put a drop of water on the polish and cover with a cover slip. Examine the slide following the instructions on page 25. You should be able to see the shape of the leaf's stomata and leaf cells imprinted in the nail lacquer. If you are not able to identify the stomata, you may wish to try again with a different type of leaf.

5. Draw what you see at high power in the space below. Be sure to label your drawing and record the magnification of the microscope.

Title: ....

Magnification: ....

Notes about specimen or procedure:

....
....
....
....
....
....
....
....

Chapter 11

Workbook 11.1

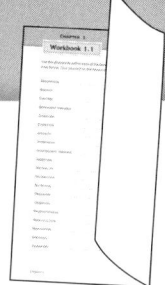
Sample

Use the glossary to define each of the keywords on the lines below. Quiz yourself on the keywords every day until you can explain the meaning of each term without looking at the definition.

ASEXUAL REPRODUCTION ....

COROLLA ....

CROSS-POLLINATION ....

EMBRYO ....

FERTILIZATION ....

GAMETES ....

GERMINATION ....

MITOSIS ....

PISTIL ....

POLLINATION ....

SEED DISPERSAL ....

SELF-POLLINATION ....

SEXUAL REPRODUCTION ....

SPORULATION ....

STAMEN ....

VEGETATIVE PROPAGATION ....

ZYGOTE ....

Fold

Fold

Workbook 11.2

A. Compose a short paragraph explaining the major difference between asexual and sexual types of reproduction. Give at least one advantage of each type.

....
....
....
....
....
....
....
....
....
....

B. Neatly write the following words and phrases in the correct order, starting with the phrase, "egg cell forms":

| egg cell forms | dispersal | fertilization | germination |
| pollination | sperm cell forms | zygote grows | |

....
....
....

Chapter 11

Workbook 11.3

Fill in the blanks with the correct keyword. Use each keyword only once.

1. Fungi and some plants reproduce asexually by ....

2. Budding is a form of .... involving only one parent.

3. Grafting, layering, and cuttings are all methods of ....

4. The .... of a flower attracts pollinators.

5. The filament and anther are parts of the ....

6. The .... is all the female parts of a flower.

7. Flowers are special organs for .... in which two cells unite.

8. The egg and sperm may be called ....

9. When the egg and sperm unite they form a ....

10. .... is when the pollen from one flower reaches the stigma of another.

11. Maple seeds have wings for .... by the wind.

12. If conditions are right, ... occurs and the embryo grows into a mature plant.

Workbook 11.4

In the blank in front of each statement, identify the type of asexual reproduction with a: B, F, S or V.

B = Budding **F** = Binary Fission
S = Sporulation **V** = Vegetative Propagation

1. .... An amoeba splits into two.

2. .... A large hydra forms a tiny hydra on its side.

3. .... A mushroom forms a cap.

4. .... Potato skins grow shoots and roots.

5. .... A yeast cell forms a tiny cell on the side.

6. .... Onions are planted from bulbs.

7. .... A seedless orange tree branch is grafted onto a normal orange tree.

8. .... A major method of reproduction for bacteria.

9. .... Horizontal shoots from strawberry plants take root.

10. .... Ferns form sori on the undersides of the fronds.

Workbook 11.5

Diagram, label, and color a complete flower in the space below. Refer to Figure 11.9 in the text. You may use a pencil to draw the diagram, but use a pen to write the labels. Use a ruler for any straight lines, and don't forget to write a title.

Title: ....

Workbook 11.6

Draw and label a diagram of the five stages of mitosis. Include at least two chromosomes in your initial cell. Refer to Figure 11.10 in the text.

Title: ....

CHECK IT OUT!

Animation of mitosis: *http://www.johnkyrk.com/mitosis.html*

For extra credit, complete the activity at the following link:
http://bio.rutgers.edu/~gb101/lab2_mitosis/section1_frames.html

Workbook 11.7

Grafting Cacti

Grafting is one method of vegetative reproduction. There are two parts of a grafted plant: the **stock**, which is attached to the ground, and the **scion**, which is the piece of another plant that is grafted onto the stock.

Supplies:
- two small, potted cacti of different varieties (For best results, choose cacti with thick, barrel-like stems.)
- sharp paring knife
- string or rubber bands
- gardening gloves

Cacti are among the easiest plants to graft. The key to grafting cacti is to make sure the cacti's vascular cambium rings overlap. As you learned in Chapter 10, a plant's cambium is the special area of cell division within the stem. In cacti, the cambium looks like a ring of dots around the center of the stem (see Figure 1). If the cambium in the scion does not touch the cambium of the stock at least a little bit, it will be very hard for stock and scion to grow together. Study Figures 1, 2, and 3 and read all the instructions carefully before grafting your cacti.

1. Obtain two small cacti of different varieties. (If the cacti are *too* different from each other, they will not grow together well. Cacti from similar but not identical species are best for grafting.)

2. Choose one cactus to be the stock and one to be the scion. If one cactus is larger and thicker than the other, it will probably make a good stock.

3. Wearing gloves to protect your hands from cactus spines, use a clean, sharp knife to slice through the stock cactus horizontally. Try to cut through the freshest and juiciest part of the cactus, but do not cut too close to the base or too close to the top. Identify the cambium ring in the stock.

4. As soon as your stock is prepared, slice one or two inches off the top of the scion. Once again, try to cut through the freshest and juiciest part of the cactus. Do not cut off too much of the scion, or it will be difficult to balance the scion on top of the stock. Identify the cambium ring in the scion.

FIGURE 1. VASCULAR CAMBIUM

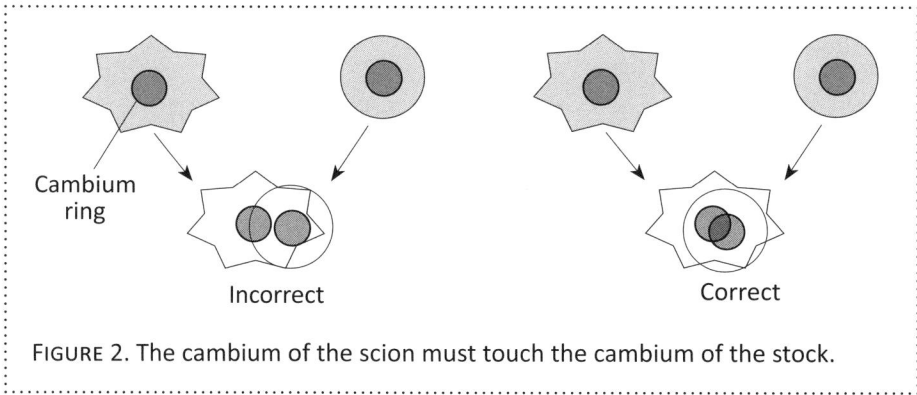

FIGURE 2. The cambium of the scion must touch the cambium of the stock.

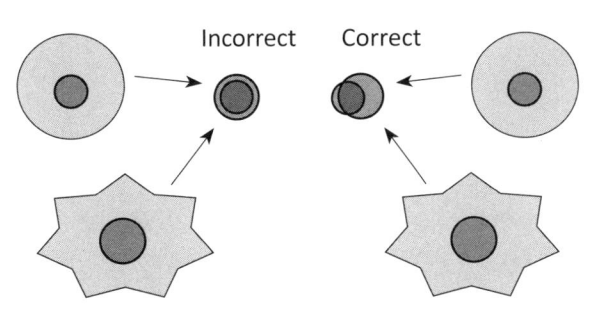

FIGURE 3. The area inside the cambium ring is not part of the cambium. If the cambium ring of one cactus is smaller than the that of the other, the rings should be placed off center so that they overlap.

5. Before joining the scion and the stock, check to make sure the top of the stock is still wet and juicy. If it has started to dry out, slice a little more off the top. Then put the scion on top of the stock, lining them up so that the cambium rings overlap, as in Figures 2 and 3.

6. Fasten the scion on top of the stock with string or rubber bands, making sure the bonds are tight enough to hold the two cacti in position, but not so tight that they tear through the skin of the cacti. (Figure 4)

7. Once your first cactus has been grafted, you may wish to graft together the top and stem which are "left over" from your first grafting. To do this, simply cut fresh slices off the two pieces of cacti and repeat steps 5 and 6.

8. Keep your cactus or cacti inside, out of direct sunlight, and keep the soil moist but not soggy. After one to three weeks, the scion and stock should be growing together as one plant.

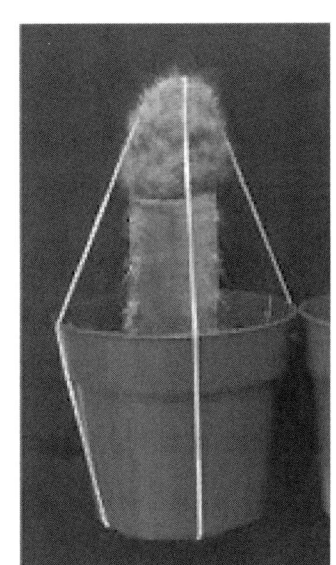

FIGURE 4.

Note: If the scion begins to dry out and turn brown, it is a sign that the grafting was not successful. Perhaps the cuts in the cacti allowed disease-causing bacteria to infect the plants, or maybe the cambium of the scion wasn't lined up correctly with the cambium of the stock. Don't worry; grafting requires practice! Feel free to try again if you wish.

Visit http://www.kadasgarden.com/grafting2.html *for more details on grafting cacti.*

Tips for growing cacti:
Do not overwater! Cacti were designed to grow in the desert, so they need good drainage. When the cacti are newly grafted, you should water regularly to keep the soil moist, but after the first few weeks, allow the soil to dry out completely between waterings.

Chapter 11

Workbook 11.8

Microscope: Cactus Spines

Supplies:
- microscope
- microscope slide and cover slip
- spines from two or more types of cacti
- tweezers

1. Use tweezers to carefully remove a spine from one cactus. Place the spine on the microscope slide and examine it under the microscope, following the instructions on page 25. Remember to begin with the low power objective lens and then proceed to the high power lens. Do not use the high power lens without using a cover slip.

2. Study the shape of the spine, and then compare with a spine from a different variety of cactus. If you look carefully, you may find that some of the spines end in tiny hooks!

Chapter 12

Workbook 12.1

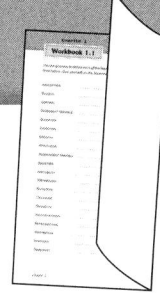

Sample

Use the glossary to define each of the keywords on the lines below. Quiz yourself on the keywords every day until you can explain the meaning of each term without looking at the definition.

ANIMALIA ....

ANNELIDA ....

ANTERIOR ....

ARTHROPODA ....

BILATERAL SYMMETRY ....

CNIDARIA ....

DORSAL ....

ECHINODERMATA ....

EXOSKELETON ....

LARVA ....

METAMORPHOSIS ....

MOLLUSCA ....

NEMATODA ....

NYMPH ....

PLATYHELMINTHES ....

PORE ....

PORIFERA ....

POSTERIOR ....

PUPA ....

RADIAL SYMMETRY ....

REGENERATION ....

SESSILE ....

TENTACLE ....

VENTRAL ....

Workbook 12.2

Use Figure 12.2 in the text to answer each question with a full sentence.

1. Are bluegills and humans in the same order?

   ....

   ....

2. An animal called the pumpkinseed is in the Genus Lepomis. Which animal from Figure 12.2 is the pumpkinseed most like?

   ....

   ....

3. A gray seal is in the same class as the bottlenose dolphin. Must it also be in the same phylum?

   ....

   ....

4. How many kinds of organisms will be in the Genus Tursiops *and* species truncatus?

   ....

   ....

5. What is the scientific name of a blue whale?

   ....

   ....

CHECK IT OUT!   Time lapse video of monarch butterfly life cycle:
http://education.nationalgeographic.com/education/activity/monarch-butterfly-life-cycle-and-migration/?ar_a=1

Workbook 12.3

Draw a diagram and write a short paragraph using complete sentences to explain complete metamorphosis. Refer to Figure 12.25 in the text. Include: diagram, title and labels.

Title: ....

Workbook 12.4

Diagram, label, and color a generalized insect in the space below. Refer to Figure 12.23 in the text. You may use a pencil to draw the diagram, but use a pen to write the labels. Use a ruler for any straight lines, and don't forget to write a title.

Title: ....

CHECK IT OUT! Virtual cockroach dissection: *http://www.orkin.com/cockroaches/virtual-roach/*

Workbook 12.5

A. *Distinguish between Kingdom Animalia and the other five kingdoms by listing 3 of the main characteristics of animals.*

1. ....
   ....

2. ....
   ....

3. ....
   ....

B. *Fill in the blanks with the correct keyword.*

1. When you are "glued to your seat" at an exciting part of the movie *The Lion, the Witch, and the Wardrobe*, you are ....

2. Sponges are full of ....

3. A .... is the stage of a butterfly just before it becomes an adult.

4. Starfish and pizza both have ....

5. Because of ...., a sponge can completely regrow from just a fragment of damaged sponge.

Workbook 12.6

The chart below organizes basic information about the creatures in each phylum of Kingdom Animalia. Study the chart carefully, and then fill in the numbered blanks, referring to the information provided about each phylum in Chapter 12. Omit Phylum Hemichordata. Since you have not yet studied Phylum Chordata, the information for that phylum has been filled in for you.

PHYLUM	1)	4)	PLATYHELMINTHES
Common name	Sponges	Jellyfish	8)
Example	2)	5)	9)
Primary characteristics	Body full of pores	Bag-like body, tentacles	Ribbon-like body
Symmetry	Not symmetrical	6)	10)
Reproduction	• Sexual: hermaphroditic • Asexual: budding and regeneration	• Sexual: hermaphroditic • Asexual: budding and regeneration	• Sexual: hermaphroditic • Asexual: regeneration
Tissue layers	No true tissues	7)	11)
Organs and systems	• No organs or systems • Skeleton made of spongin	• Sensitive skin with nerve net • Gastrovascular (digestive) cavity	• Simple nervous system with brain and eyespots • Dead-end digestive system
Relation to humans	3)	Jellyfish can sting you.	Some species are parasites; others are decomposers.

Continued on next page.

PHYLUM	12)	Annelida	18)
Common name	Roundworms	15)	Mollusks
Example	13)	Earthworms, leeches	19)
Primary characteristics	Thread-like body	Many segments or rings	20)
Symmetry	Bilateral symmetry	Bilateral symmetry	Bilateral symmetry
Reproduction	• Sexual: some species are hermaphroditic; most have distinct sexes • Asexual: none	• Sexual: 16)......................... • Asexual: regeneration in some species	• Sexual: some species are hermaphroditic; most have distinct sexes • Asexual: none
Tissue layers	Three tissue layers	Three tissue layers	Three tissue layers
Organs and systems	• Nervous system with two main nerves • 14)	• Nervous system with simple brain • Complete digestive system • Respiration: 17)........................	• Nervous system with two main nerves • Definite head and sense organs • Complete digestive system • Gills
Relation to humans	Some species are parasites; others are decomposers.	Some species are parasites, others are decomposers or help to aerate the soil.	Many species are eaten for food. Some are garden pests; others are disease vectors.

Continued on next page.

PHYLUM	21)	25)	Chordata
Common name	Arthropods	Starfish	Chordates
Example	22)	26)	Fish, humans, birds
Primary characteristics	23)	27)	Notochord and dorsal nerve cord
Symmetry	Bilateral symmetry	28)	Bilateral symmetry
Reproduction	• Sexual: distinct male and female sexes • Asexual: none	• Sexual: distinct male and female sexes • Asexual: 29)............................	• Sexual: distinct male and female sexes • Asexual: none
Tissue layers	Three tissue layers	Three tissue layers	Three tissue layers
Organs and systems	• Nervous system with simple brain • Distinct head and variety of sense organs • Complete digestive system • Exoskeleton • Specialized respiratory organs such as gills or air tubes	• No centralized nervous system • Usually complete digestive system • Internal skeleton covered by thin epidermis	• Complex nervous system • Complete digestive system • Internal skeleton • Respiratory system
Relation to humans	Some species are parasites; others are decomposers. 24)	Some species are pests.	Some species are pets or domesticated animals; some are eaten for food; some are pests.

Chapter 12

Workbook 12.7

Animal Dissection

> **Supplies:**
> - preserved animal for dissection
> - household cutlery or dissection kit

Dissect a fish, frog, shark, starfish, squid, or other preserved specimen from Kingdom Animalia. Most of these can be purchased as preserved specimens for less than $10 from supply companies such as *www.enasco.com* or *www.carolina.com*. Household cutlery can be used for the dissection, but it may be worthwhile to invest in a basic dissection kit, since the student will probably be doing more dissections in high school.

Detailed instructions and videos for dissecting each type of animal can be found online. *Do not* simply cut into the animal at random, or you may damage many of the organs which you wish to study.

Chapter 13

Workbook 13.1

Use the glossary to define each of the keywords on the lines below. Quiz yourself on the keywords every day until you can explain the meaning of each term without looking at the definition.

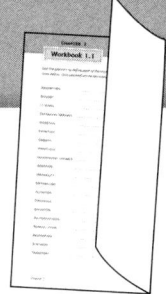

Sample

Agnatha ....

Amphibia ....

Aves ....

Carnivore ....

Chondrichthyes ....

Chordata ....

Cold-Blooded ....

Estivate ....

Gills (2) ....

Herbivore ....

Hibernate ....

Insectivore ....

Lungs ....

Mammalia ....

Omnivore ....

Osteichthyes ....

Placenta ....

Reptilia ....

Uterus ....

Vertebrata ....

Warm-Blooded ....

Workbook 13.2

Fill in the blanks with the correct keyword.

1. A person who eats a Whopper hamburger with lettuce, tomato, pickles, and all the toppings is an ....

2. When St. John the Baptist lived in the desert on locusts (Class Insecta) he was an ....

3. A dorsal nerve cord is a defining characteristic of Phylum ....

4. Young amphibians breathe underwater by means of ....

5. On a cool night, snakes and other reptiles slow down because they are ....

6. Your belly button is where your umbilical cord once attached you to your ....

7. The mistaken theory that frogs can form spontaneously out of mud may have been caused by frogs' ability to ....

CHECK IT OUT! Video of anglerfish catching minnow:
http://www.arkive.org/anglerfish/lophius-piscatorius/video-08.html

Workbook 13.3

Write a short paragraph in which you name and describe the special organ you used when living inside your mother before birth. Name three other examples of animals whose young use the same kind of organ. Name at least one mammal that does not use this organ to live inside its mother.

Workbook 13.4

Diagram, label, and color an animal cell in the space below. Refer to Figure 5.8 in the text. You may use a pencil to draw the diagram, but use a pen to write the labels. Use a ruler for any straight lines, and don't forget to write a title.

Title: ....

Workbook 13.5

Using a complete sentence, answer each question by using the words below.

Agnatha	Chondrichthyes	Nematoda
Amphibia	Chordata	Osteichthyes
Animalia	Cnidaria	Platyhelminthes
Annelida	Echinodermata	Porifera
Arthropoda	Mammalia	Reptilia
Aves	Mollusca	Vertebrata

1. Which terms are names of kingdoms?

   ....
   ....

2. Which are names of phyla?

   ....
   ....

3. Which are names of subphyla?

   ....
   ....

4. Which are classes?

   ....
   ....

5. Which group includes all the rest?

   ....
   ....

Workbook 13.6

Fill in the blanks with the phylum name most associated with the term or phrase.

Phylum Annelida	Phylum Echinodermata	or use:
Phylum Arthropoda	Phylum Mollusca	All of these phyla
Phylum Chordata	Phylum Nematoda	None of these phyla
Phylum Cnidaria	Phylum Platyhelminthes
Phylum Porifera

1. Animals in .... are pore-bearers.

2. Animals in .... have skeletons of spongin.

3. Animals in .... are heterotrophic.

4. Animals in .... have cell walls made of chitin.

5. Animals in .... have sac-like bodies.

6. The head of the tapeworm, an animal in ...., is a scolex.

7. .... contains organisms in Class Angiospermae.

8. Animals in .... carry out respiration.

9. Animals in .... are also called round worms.

10. Animals in .... have chloroplasts.

11. Segmented worms are in ....

12. Animals in .... are important in forming good soil.

13. Animals in .... lack nuclear membranes.

14. Animals in .... are soft bodied with a muscular foot.

15. Animals in .... are made of cells.

16. Animals in .... have a chitinous exoskeleton.

Chapter 13

17. .... contains joint legged animals.

18. .... contains spiny skinned animals.

19. .... contains creatures with a dorsal nerve cord.

20. Insects are in ....

21. Birds are in ....

22. Snakes are in ....

23. Frogs are in ....

24. Sponges are in ....

25. Fish are in ....

26. .... contains Class Crustacea.

27. Animals in .... reproduce.

28. Spiders are in ....

29. Amoeba sp. are in ....

30. Monocots are in ....

31. Mammals are in ....

32. Jellyfish are in ....

33. Planaria sp. are in ....

34. Hookworms are in ....

35. Lumbricus sp. are in ....

36. Butterflies are in ....

37. Sharks are in ....

38. Placentals are in ....

39. Homo sapiens is in ....

Workbook 13.7

Microscope: Feather

Supplies:
- microscope
- microscope slide and cover slip
- feather
- eyedropper
- water

1. Place the feather on a slide. Examine the microscope slide without a cover slip with the microscope on low power.

2. To view the feather on high power, cut off a small portion from the thinnest area of the feather and place it on a slide. Add a drop of water and a cover slip. View slide under the microscope, following the instructions on page 25. You will be able to see the delicate strength of the tiny, overlapping "hairs" (called barbules) that hold the feather together.

3. Draw what you see at high power in the space below. Be sure to label your drawing and record the magnification of the microscope.

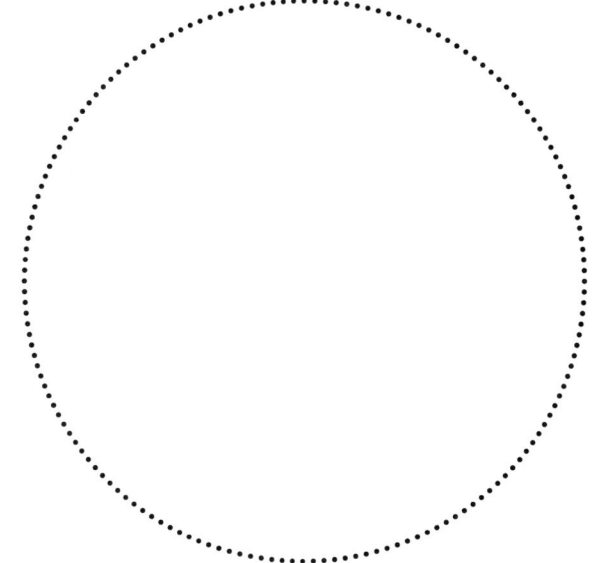

Title: ....

Magnification: ....

Notes about specimen or procedure:

....
....
....
....
....
....
....
....
....

Chapter 13

Midway Review

The questions in the Midway Review touch on the major topics you have studied so far. Parts II-V of the Midway Review are designed to be given "closed-book." So that you will be prepared to do your best, review your text carefully, paying special attention to key terms, chemical equations, and charts. You should know the meaning and correct spelling of all the keywords in Chapters 1-13.

Part I. Keywords

A. **Spelling**

1. After you have studied the spelling and definitions of the keywords from Chapters 1-13, ask someone to test you orally on each word using the keyword and definition lists in the workbook. When he reads the definition, correctly spell the keyword on a separate sheet of paper.
2. Restudy any of the keywords you missed and repeat step 1 until you know them all.

B. **Definitions**

1. Now have someone test you on the definitions. When he says the keyword, correctly explain its meaning.
2. Restudy any of the keywords you missed and repeat step 1 until you know them all.

Midway Review

Part II. Measurements and Chemistry

1. List the six steps of the scientific method.

   ....

   ....

   ....

   ....

   ....

   ....

2. Convert:

   5 m = ... cm   847 ml = ... l

   8 cc = ... ml

3. Correctly spell the name of each element:

   C: ....   H: ....   O: ....

   N: ....   S: ....   P: ....

   I: ....   Na: ....   K: ....

   Fe: ....   Cu: ....   Cl: ....

   Mg: ....   Ca: ....   Zn: ....

4. Write the formulas for the following compounds:

   Water: ....

   Oxygen gas: ....

   Table salt: ....

   Carbon dioxide: ....

Midway Review

Part III. Cells

1. List the eight life functions.

   ....

   ....

   ....

2. Briefly describe the primary function of each cell part listed below.

   Cell wall: ....

   Centrioles: ....

   Chloroplast: ....

   Chromosomes: ....

   Endoplasmic reticulum: ....

   Golgi body: ....

   Lysosome: ....

   Mitochondrion: ....

   Nucleus: ....

   Ribosomes: ....

   Vacuoles: ....

3. Explain the process of osmosis.

   ....

   ....

   ....

   ....

   ....

Midway Review

Part IV. Classification

1. List the seven taxons in order from largest to smallest.

   ....
   ....
   ....
   ....

2. List the six kingdoms.

   ....
   ....
   ....

3. What are the three basic shapes of bacteria?

   ....
   ....

4. List the three main phyla of plants along with an example of a plant in each phylum.

   ....
   ....
   ....

5. List three classes of Phylum Tracheophyta with examples.

   ....
   ....
   ....

Midway Review

6. Name the two subclasses of Class Angiospermae and describe at least three of the differences between plants in the two subclasses.

   ....
   ....
   ....
   ....
   ....

7. List the ten main phyla in Kingdom Animalia, with an example of each.

   ....
   ....
   ....
   ....
   ....
   ....
   ....
   ....
   ....
   ....

8. Identify the main characteristics of each of the seven classes in Phylum Chordata.

   ....
   ....
   ....
   ....
   ....
   ....
   ....

Midway Review

Part V. Reproduction, Growth, and Response

1. Identify four types of asexual reproduction.

   ....

   ....

   ....

   ....

2. What is the process of mitosis? Include the names of the five stages in the process.

   ....

   ....

   ....

   ....

   ....

3. Explain the main differences between sexual and asexual reproduction. Identify at least one advantage to each method of reproduction.

   ....

   ....

   ....

   ....

4. Explain how phototropism works.

   ....

   ....

   ....

   ....

Animal Classification Research Paper

The goal of this assignment is to learn about and write an informative paper on an animal of your choice, including information about the characteristics that cause it to be classified as it is. (For example, Phylum Chordata, Subphylum Vertebrata, Class Mammalia, etc.) Be sure to outline your topic before writing your first draft.

The paper will include the following information about your selected animal:

- physical characteristics (including information about why the animal is classified as it is)
- environment, and characteristics that enable the animal to thrive therein
- food and feeding habits
- camouflage and protective behaviors
- care and feeding of young
- any especially interesting, unusual, or important details about your animal

Length of text

Your paper should be double-spaced and run at least 650–750 words, not including the Table of Contents, diagrams/charts, glossary, and Works Cited.

Format

Follow directions given in your English text, a text of your choice, or in the "Tips for Writing a Research Paper" (pages 143–146).

What else to include in your paper:

- Use a minimum of five vocabulary words from Chapters 12 or 13 in your paper.
- Include a Works Cited page with a minimum of three different sources. (These sources might include library books, magazine articles, the encyclopedia, and one internet article.)
- Include at least three quotations from your sources. See "Tips for Writing a Research Paper" on pages 143–146 for instructions on formatting quotations.
- Include in your paper a description of the animal that shows why it is classified into each of its taxons.

- Provide a diagram, map, graph, or chart related to your topic. (For example, in addition to the written description, the paper might include a diagram showing why the animal is classified into each of its taxons.) Be sure to refer to this page in your report!

- Include a sheet with full classification of your animal.

- Provide a glossary defining the minimum of five words from this chapter that were used in your paper.

Putting It All Together

Your paper should be assembled in this order:

- **Cover** with artwork related to the topic *(optional)*

- **Title page**

- **Table of Contents** listing the following parts of the paper along with the page number on which the information begins

- **Text** of the paper

- **Full classification** of your animal on a separate sheet of paper

- **Diagram, map, graph, or chart** related to your topic and referenced in your report

- **Glossary**

- **Works Cited** page

Please use the information above as a check list. Consult it as you write your paper to be sure that you have included all required information.

Tips for Writing a Research Paper

Purpose

Research papers are written as much to inform the writer as they are to inform the reader. The author's goal, then, should be to gather and research from as many different sources as time allows in order to harvest the greatest amount of information. This collection of information enables the author to develop a better understanding of his subject and, from the wealth of information, makes it easier for him to write his paper.

First, with your teacher's assistance, set up a due date and a schedule of time allotted for research and note-taking; writing a rough draft; and writing a final draft. Consult with your teacher to see if you might also receive English credit for this paper, or if it might be substituted for another English writing assignment.

Format

Research papers may be written and formatted in a number of different styles. In time, you will learn the latest (and seemingly ever-changing) style accepted by the Modern Language Association (MLA) but, for now, you may use this simplified format for your paper.

In addition, please reference instructions for essay or report-writing employed by your current English book. If the instructions in your book are clear, please use the format suggested there. However, if your text contains no directions for writing essays or reports, you may wish to obtain a copy of *Language of God Level E*, *F*, or *G*; or *Jensen's Format Writing* (all available from CHC). Each of these books contains guidelines for writing essays and papers.

Following is a suggested sample format for an essay on the North American pika. (This is simply a sample format and not one which you must follow exactly.)

Title page for your paper

Title pages contain:

> paper's title
> your name
> name of class
> date

Table of Contents

The table of contents lists the parts of the paper along with the page number on which the information begins. You do not need to include the title page or the table of contents in this list. For instance, your table of contents might look like this:

> Elusive Mountain-Dweller: the North American Pika, page 2
> Taxonomy of Ochotona princeps, page 5
> Map of the Pika's Distribution, page 6
> Glossary, page 7
> Works Cited, page 8

Sample Outline of Text:

Elusive Mountain-Dweller: the North American Pika

I. Introduction *(One or two paragraphs in which each of the paper's topics is mentioned briefly. In this paper, those topics are listed in headings II, III, IV, and V, below.)*

II. Anatomy, Physical Characteristics, Classification (use reference and quote from book)
- A. size
- B. fur and color
- C. limbs, fur, herbivore's teeth
- D. related to rabbits
- E. reference taxon chart here—Chordata, Vertebrata, Mammalia, etc.

III. Feeding Habits
- A. herbivore
  - grasses, thistles, other green plants
- B. water requirements
  - drink water, but also can take moisture primarily from vegetation

IV. Environment and Characteristics Suited to the Environment (use reference and quote from web article)
- A. mountain dweller
  1. above tree line
  2. create "caves" in loose rock (use quotation from "The Invisible Life of the Pika" here)
- B. fur
  - seasonal changes in thickness and color
- C. cope with temperature changes by retreating to rock caves, which are cooler in summer and warmer in winter
- D. "harvest" grasses, store them in piles near burrows for winter consumption

V. Protective Behaviors and Camouflage
- A. fur color helps camouflage against rocks
- B. shrill "whistling" warning call

VI. Pika Breeding and Young (reference and quote from magazine article)
- A. animals breed about one month before snowmelt
- B. gestation is 30 days (mention God's good design to coincide with snowmelt when more food will be available)
- C. two litters per year, about three babies each
- D. born with a mouthful of teeth; little fur; eyes open at nine days
- E. weaned at about a month and able to forage for themselves

VII. Conclusion *(One or two paragraphs in which each of the paper's topics is again mentioned briefly, with a sense of concluding. Use transition words like "therefore," "hence," "thus," "finally," and "in conclusion.")*

Glossary

Glossaries are similar to dictionaries in that they include "vocabulary words" introduced in the paper and their definitions. Your glossary will appear just before the Works Cited page at the end of the paper.

Works Cited page

This page appears last in your paper. On this page are listed all the books, magazines, articles, and websites from which you quoted in your paper. There are variations of presentation but, for the purpose of this paper, you will use a basic format. "Medium of publication" indicates whether it is a book (Print), magazine article (Print), or from the web (Web).

citing a book:

Author's last name, first name. *Book Title*. City where book was published: Publisher, year published. Medium of publication.

Example: Bowling, I.M. *Hairy Petters*. London: Flights of Fancy, 2013. Print.

citing an article in a reference book:

"Title of Article." *Title of Reference Work*. Edition. Year. Medium.

Example: "Small Animals I Have Known and Loved." *The Encyclopedia of Tiny*. 2nd ed. 1997. Print.

citing a magazine article:

Author's last name, first name. "Article Title." *Periodical Title* day month year: page(s). Medium of publication.

Example: Shakefear, William. "Taming of the Pika." *Mountain Critters Quarterly* 2 Feb. 2013: 16. Print.

citing a web article:

Author's last name, first name. "Article Title." *Name of Website*. Name of Organization. Article date. Web. Date you read the article.

Example: Unseene, Carolyn. "The Invisible Life of the Pika." *MountainMysteries*. Animal Detectives. 2 Feb. 2013. Web. 31st Feb 2013.

Tips for Writing a Research Paper

References are listed alphabetically on the Works Cited page, by the author's last name. If there is no author, go right to the next piece of information in the citation to determine alphabetical placement. (From author to title of book or article, etc.)

Citing Quotations

When quoting an author, use his/her last name, and the page number from which the quotation was drawn. For example,

> Truman once remarked, "In reading the lives of great men, I found that the first victory they won was over themselves . . . self-discipline with all of them came first" (123).

Notice that the page number follows the quote, followed by the punctuation.

On the Works Cited page, the quotation would be referenced thus:

> Truman, Harry S. *Book Title*. Place of publishing: Name of Publisher, year of publishing. Medium.

Chapter 14

Workbook 14.1

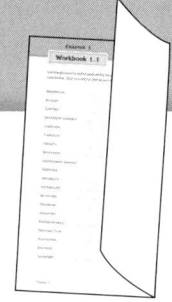
Sample

Use the glossary to define each of the keywords on the lines below. Quiz yourself on the keywords every day until you can explain the meaning of each term without looking at the definition.

Amino Acids ....

Carbohydrate ....

Cellular Respiration ....

Chemosynthesis ....

Disaccharide ....

Enzymes ....

Fat ....

Glucose ....

Glycogen ....

Mineral ....

Monosaccharide ....

Nutrient ....

Polysaccharide ....

Protein ....

Vitamin ....

Workbook 14.2

Write a keyword from Chapter 14 to fill in each blank.

1. The human body uses .... for lubrication, insulation, and long term storage of energy.

2. Proteins are made of nitrogen-containing pieces called ....

3. $C_6H_{12}O_6$ is the formula for ....

4. The general name for any simple sugar made of just one sugar ring is ....

5. Wood is not a .... for you because humans cannot digest wood.

6. .... are long chains of amino acids.

7. Photosynthesis is the opposite of ....

8. Specialized bacteria use .... to turn hydrogen sulfide into sulfur.

Workbook 14.3

If necessary change the italicized word to make the statement true. Write each full statement neatly once it is true.

1. Hydrogen oxide is another name for *starch*.

   ....
   ....

2. A few types of *amino acids* make up thousands of kinds of proteins.

   ....
   ....

3. *Iodine* is the mineral in hemoglobin that carries oxygen.

   ....
   ....

4. *Proteins* are made of carbon, oxygen, and hydrogen, with two hydrogens for every oxygen.

   ....
   ....

5. Sucrose is a *monosaccharide*.

   ....
   ....

6. Glycogen is a *starch* made from glucose molecules.

   ....
   ....

7. Sodium and chlorine are *minerals* in table salt.

   ....
   ....

8. *Ascorbic acid* is another name for vitamin C.

   ....
   ....

9. *Thiamine* is another name for the vitamin A in carrots.

　　..
　　..

10. Milk and milk products are a good source of *iron*, a mineral needed for strong bones.

　　..
　　..

11. ATP is a molecule used to hold *energy*.

　　..
　　..

12. *Glucose* is combined with oxygen in cellular respiration.

　　..
　　..

13. Chlorophyll is needed for *respiration*.

　　..
　　..

14. In photosynthesis, *carbon dioxide* is given off as a waste.

　　..
　　..

15. Human beings are *autotrophs*.

　　..
　　..

Workbook 14.4

A. List the five nutrient groups and the name of a food from that nutrient group.

Nutrient Group	Food
....	....
....	....
....	....
....	....
....	....
....	....

B. Write the chemical equation for cellular respiration.

....

....

C. Write the chemical equation for photosynthesis.

....

....

Workbook 14.5

Compose a paragraph comparing and contrasting how a green corn plant and a cow obtain energy. Correctly use at least five of the following terms:

ATP	heterotroph	autotroph	monosaccharide
chlorophyll	photosynthesis	glucose	cellular respiration

Workbook 14.6

Fill in the blanks with the correct vitamin or mineral. Terms may be used more than once.

Vitamin A	Vitamin B_1	Vitamin C	Vitamin D	Iron (Fe)
Calcium (Ca)	Iodine (I)	Sodium (Na)	None of these	

1. .... is also called ascorbic acid.

2. .... builds strong bones and teeth.

3. .... helps control heartbeat.

4. .... is found in citrus fruits such as limes.

5. .... is in hemoglobin.

6. .... is needed for the thyroid gland.

7. .... helps nerves work properly.

8. .... is in table salt.

9. .... is a disaccharide.

10. .... is a fat.

Workbook 14.7

In the space provided, identify which nutrient group is most closely associated with the statement. Use the letters below.

C = carbohydrates   F = fats and oils
P = proteins   V = vitamins and minerals
W = water   X = none of these

1. .... Glucose and fructose are simple sugars.

2. .... This nutrient group makes up 60% of your body.

3. .... Hair and fingernails contain nitrogen.

4. .... A molecule of cellulose contains two hydrogen atoms for every oxygen atom.

5. .... Meats are made of amino acids.

6. .... A high fever can destroy human enzymes.

7. .... Carotene helps the eyes.

8. .... Glycogen is a starch.

9. .... Meats, eggs, and milk are necessary to provide amino acids which your body cannot make.

10. .... Enzymes help chemical reactions to take place at low temperatures.

Workbook 14.8

Identify each of the nutritional groups pictured. Write a complete sentence to describe each group.

Example

Vegetables are a good source of vitamins.

A.

B.

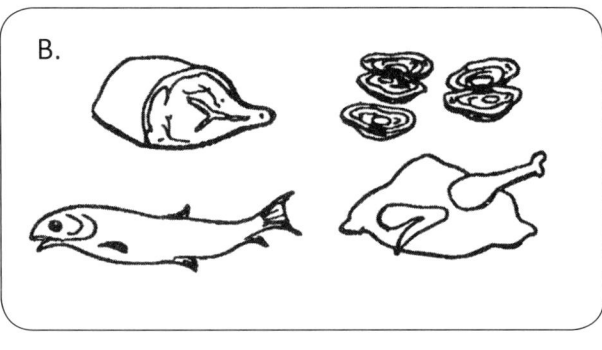

C.

D.

14.9 Formal Lab #4

Complete the experiment below, following the instructions provided. Fill in the blanks as you go. Use complete sentences to answer the questions at the end.

Name: .... Date: ....

I. Title: Yeast and Sugar

II. Purpose:

Yeast is a fungus in Phylum Ascomycota. Like all fungi, yeast is heterotrophic, which means it cannot produce its own energy. Instead, yeast gets its energy from nutrients digested outside of the cells.

The species of yeast used in making bread, <u>Saccharomyces cerevisiae</u>, gets its energy by digesting sugar. At the same time, the yeast releases the waste product carbon dioxide.

In this experiment we will attempt to discover whether the amount of carbon dioxide released by <u>Saccharomyces cerevisiae</u> depends on the amount of sugar the yeast is digesting.

My hypothesis is that the amount of carbon dioxide released by <u>Saccharomyces cerevisiae</u> .... [will/will not] increase if we provide the yeast with more sugar to use as food.

III. Materials: The materials required for this experiment are active dry yeast, water, and sugar.

IV. Apparatus: The equipment required for this experiment is five identical, empty water bottles with tight-fitting caps, five standard balloons of the same size, five rubber bands, a permanent marker, measuring spoons, a sheet of paper, a piece of string (about 15 inches long), and a centimeter ruler.

V. Procedure:

1. Label the five water bottles A, B, C, D, and E with a permanent marker. Put one cup of warm water into each water bottle. Make sure the water is warm, not hot (hot water will kill the yeast cells), and make sure the water in each bottle is at the same temperature.
2. Shape the piece of paper into a funnel, and use it to add 2 tsp. of active dry yeast to each water bottle.
3. Add ½ tsp sugar to bottle B, add 1 tsp sugar to bottle C, add 1½ tsp sugar to bottle D, and add 2 tsp sugar to bottle E. Do not add any sugar to bottle A; this bottle will be your **control**. Screw the caps onto the bottles and shake each bottle vigorously to mix the yeast and sugar with the water.
4. Remove the caps from the bottles and fit an empty balloon onto the top of each one. Wrap a rubber band tightly around the neck of each bottle to seal the balloons onto the bottles. As the yeast cells carry out cellular respiration, they will give off carbon dioxide as a waste product.

The carbon dioxide will be collected in the balloon. By measuring the size of the balloon, we can measure which bottle of yeast produced the most carbon dioxide.

5. Set the bottles in a warm place where they will not be disturbed. After half an hour, measure the circumferences of the balloons with the piece of string. Use the ruler to measure the length of the string, and record your results in Figure 1. Don't forget to include units in centimeters.
6. Measure the circumferences of the balloons every half hour until all the balloons have stopped expanding, approximately 4–6 hours. Record your results in Figure 1.
7. Use the data in Figure 1 to fill out Figure 2. For instance, if Bottle B stopped expanding at 15 cm, then write "15 cm" in Figure 2 in the space for Bottle B.
8. Use the data in Figure 2 to graph the results of your experiment in Figure 3. For instance, if you wrote "15 cm" in Figure 2 in the space for Bottle B, you would draw an X on the vertical line labeled "0.5" and slightly below the horizontal line labeled "16." Refer to pages 7–8 in the text for more information on drawing graphs.
9. After you have recorded your data in Figure 3, connect the points with a smooth curve of best fit. Complete your graph with a descriptive title such as "Circumference of balloons compared to amount of sugar in bottle."

VI. Data:

	Bottle A (no sugar)	Bottle B (0.5 tsp sugar)	Bottle C (1 tsp sugar)	Bottle D (1.5 tsp sugar)	Bottle E (2 tsp sugar)
Circumference after 0.5 hr					
Circumference after 1 hr					
Circumference after 1.5 hr					
Circumference after 2 hr					
Circumference after 2.5 hr					
Circumference after 3 hr					

FIGURE 1. CIRCUMFERENCE OF BALLOONS OVER SEVERAL HOURS

	Greatest Circumference Reached
Bottle A (no sugar)	
Bottle B (0.5 tsp sugar)	
Bottle C (1 tsp sugar)	
Bottle D (1.5 tsp sugar)	
Bottle E (2 tsp sugar)	

FIGURE 2. GREATEST CIRCUMFERENCE REACHED FOR EACH BALLOON

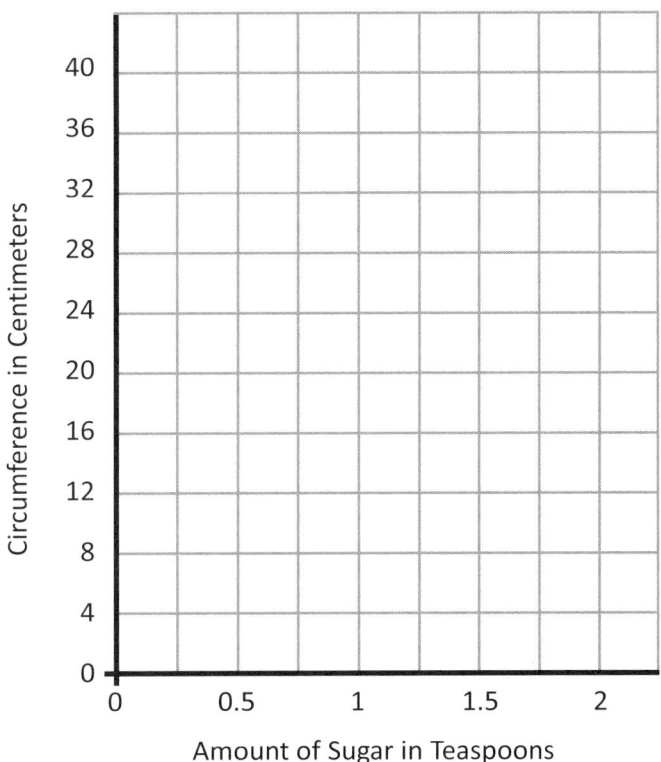

FIGURE 3. ....

Chapter 14

VII. Questions:

1. Does the amount of carbon dioxide released by <u>Saccharomyces cerevisiae</u> depend on the amount of sugar the yeast is digesting? In other words, does the production of carbon dioxide increase when we increase the sugar?

   ....
   ....
   ....
   ....

2. The **dependent variable** in this experiment is the **amount of carbon dioxide produced**. What is the **independent variable**? In other words, what does the dependent variable—the amount of carbon dioxide—depend upon?

   ....
   ....
   ....
   ....

VIII. Conclusion: This experiment showed that the amount of carbon dioxide released by <u>Saccharomyces cerevisiae</u> .... [does/does not] increase if we provide the yeast with more sugar to use as food.

Thus, my hypothesis was .... [correct/incorrect].

Chapter 15

Workbook 15.1

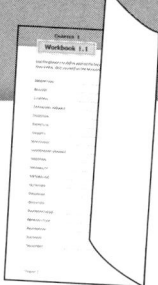

Sample

Use the glossary to define each of the keywords on the lines below. Quiz yourself on the keywords every day until you can explain the meaning of each term without looking at the definition.

BENEDICT'S SOLUTION ....

CALORIE ....

CALORIMETER ....

FILTRATE ....

HEAT ....

LUGOL'S SOLUTION ....

PRECIPITATE ....

RESIDUE ....

SOLUBLE ....

SOLUTE ....

SOLVENT ....

TEMPERATURE ....

TRANSLUCENT ....

Workbook 15.2

Do the calculations to fill in the missing parts of the chart at the bottom of the page. Use the formula for heat, as shown in the sample problem and solution below.

Formula for heat: Heat (cal) = (mass of H_2O) × ($1 \frac{cal}{g \times °C}$) × (temp. change of H_2O)

SAMPLE PROBLEM:

Heat (cal)	Mass (grams)	Initial Temp.	Final Temp.	Temp. Change
	10 g	12 °C	14 °C	

First determine the change in temperature:

Temp. Change = Final Temp. − Initial Temp.

Temperature change = 14 °C − 12 °C

Temperature change = 2 °C

Now use the formula for heat:

Heat = 10 g × $1 \frac{cal}{g \times °C}$ × 2 °C

Note: g and °C cancel

Heat = 20 calories

	Heat (cal)	Mass (grams)	Initial Temp.	Final Temp.	Temp. Change
Sample	20 cal	10 g	12 °C	14 °C	2 °C
A	1.	5 g	6 °C	7 °C	2. ....
B	3.	100 g	40 °C	43 °C	4. ....
C	5.	3 g	15 °C	6.	2 °C
D	4 cal	2 g	7.	10 °C	8. ....
E	50 cal	9.	10 °C	35 °C	10. ....

Chapter 15

163

Workbook 15.3

Use what you learned about percentages in Chapter 15 to solve the following problems.

1. If a slice of watermelon has a mass of 25 g and after drying has a mass of 2 g, what is the percentage of water in the melon?

2. If a student gets three wrong out of 20 questions, what percent of the questions did the student answer correctly?

3. If a person has a mass of 40 kg, what is the mass of the water in his body? (Hint: the human body is 60% water.)

Workbook 15.4

Using complete sentences, explain how to test for each of the following nutrients. Include the correct name of any apparatus or solutions, and give the meaning of any color changes.

1. Fat

   ....
   ....
   ....

2. Protein

   ....
   ....
   ....

3. Chlorine

   ....
   ....
   ....

4. Water

   ....
   ....
   ....

Workbook 15.5

Test for Starch

Supplies:
- cracker
- piece of cheese
- slice of apple
- pat of butter
- peanut
- Lugol's solution or tincture of iodine
- eyedropper

1. Use the eyedropper to place a drop of Lugol's solution or tincture of iodine on each food sample. Record below whether the solution remains brown or turns blue-black.

2. Refer to page 140 in the text to interpret your results and determine the starch content of each food. Record your results below.

	Color	Starch Content
Cracker		
Cheese		
Apple		
Butter		
Peanut		

Workbook 15.6

Test for Simple Sugars

Supplies:

- a small amount of:
    - meat, cooked or raw
    - bread
    - milk
    - carrot (or another vegetable)
    - apple (or another fruit)
    - table sugar
- Benedict's solution
- saucepan
- water
- one or more glass test tubes

1. Boil 4 inches of water in the saucepan. Turn the heat down to a simmer.

2. Place a small amount of the food sample at the bottom of the test tube. Cover the food with Benedict's solution and heat the test tube by holding the test tube in the pot of simmering water. You may need to use a test tube holder or potholder to hold the test tube. Do not allow the test tube to touch the bottom of the pot.

3. After 2–4 minutes, remove the test tube from the hot water bath and record the color of the solution. Use the color to determine the simple sugar content of the food, referring to Figure 15.6 on page 141 of the text. Record your results below.

4. Allow the test tube to cool, then wash it thoroughly. Repeat steps 2 and 3 with the next food sample.

	Color	Simple Sugar Content
Meat		
Bread		
Milk		
Carrot		
Apple		
Sugar		

Chapter 16

Workbook 16.1

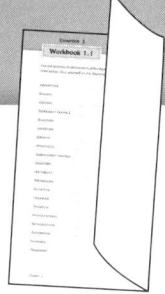

Sample

Use the glossary to define each of the keywords on the lines below. Quiz yourself on the keywords every day until you can explain the meaning of each term without looking at the definition.

Don't forget to review your keywords from Chapters 1-15! Frequent review of keywords from previous chapters is critical to doing well on the Final Review.

EXCRETION ....

GROWTH ....

IRRITABILITY ....

NUTRITION ....

ORGAN ....

REPRODUCTION ....

RESPIRATION ....

SYNTHESIS ....

SYSTEM ....

TISSUE ....

TRANSPORT ....

Workbook 16.2

Use the following terms in complete sentences to answer each question.

organelle   cell   tissue

organ   system   organism

1. Which term includes all the others?

   ....
   ....

2. Which term is the unit of structure and function in living things?

   ....
   ....
   ....

3. Which is a group of organs working together to carry out a life function?

   ....
   ....
   ....

4. Epidermis, nerves, and muscles are examples of which term?

   ....
   ....
   ....

5. Together, the mouth, esophagus, stomach, intestines, and associated organs are an example of which term?

   ....
   ....
   ....

Workbook 16.3

List each of the eight life functions except growth, and write the name of a system of the body that carries out that function.

LIFE FUNCTION	SYSTEM OF THE BODY
....	....
....	....
....	....
....	....
....	....
....	....
....	....

Workbook 16.4

Draw and color an outline of a human body and label each major region of the body. Refer to Figure 16.4 in the text. You may use a pencil to draw the diagram, but use a pen to write the labels. Use a ruler for any straight lines, and don't forget to include a title.

Title: ....

Chapter 16

Workbook 16.5

Match each term with a diagram below.

1. .... Molecule
2. .... Organelles
3. .... Cell
4. .... Tissue
5. .... Organ
6. .... System
7. .... Organism

A.

B.

C.

D.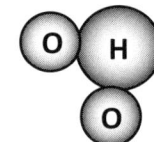

E.

F.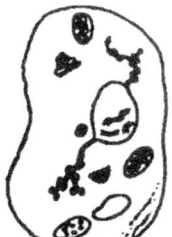

G.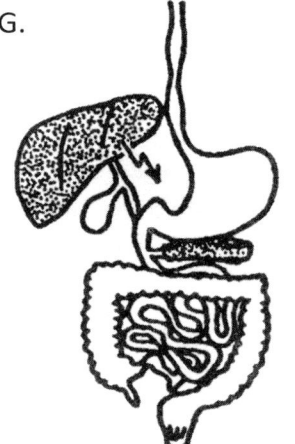

174 Chapter 16

Workbook 16.6

On a separate sheet of paper, draw a timeline of the significant events in a new life from fertilization to birth. Using a ruler, draw a line 20 cm long in the center of the page (the page must be turned sideways). Mark the beginning of life at fertilization on the left side of the timeline. End with birth on the right-hand side of the sheet, 20 cm from the beginning.

Use the data in Chapter 16 to measure out and label the significant events in a new life. To portray the periods of time described in Chapter 16, make each centimeter represent 14 days. Use the ratio $\frac{1 \text{ cm}}{14 \text{ days}}$ to convert the number of days into centimeters. For example,

$$\frac{1 \text{ cm}}{14 \text{ days}} = \frac{x \text{ cm}}{21 \text{ days}}$$

Cross multiply, cancel out units, and solve: 21 days = 1.5 cm

Workbook 16.7

Sanctity of Human Life Report

Learn about a person who has overcome a physical handicap. How does this person's life reveal the value of every human life, regardless of handicaps? Ideas: Bl. Hermann the Cripple, Nick Vujicic, and Gianna Jessen. Share your findings orally with your family.

Workbook 16.8

Microscope: Cheek Cells

Supplies:
- microscope
- microscope slide and cover slip
- tincture of iodine or Lugol's solution
- toothpick
- eyedropper

1. Gently scrape the inside of your cheek with a toothpick. Draw the toothpick across a microscope slide to deposit cheek cells on the slide. Wait for the small amount of moisture to dry, then observe the slide without a cover slip with the microscope on low power. The cells are transparent, so in order to see them you must close the microscope's diaphragm so that only a small amount of light reaches the slide.

2. Now stain the cells with tincture of iodine or Lugol's solution. Place one drop of iodine or Lugol's solution over the cheek cells on your dry slide. Lower a cover slip onto the slide and observe the slide under the microscope, following the instructions on page 25. The cells and their nuclei should now be much easier to see.

3. Draw what you see at high power in the space below. Be sure to label your drawing and record the magnification of the microscope.

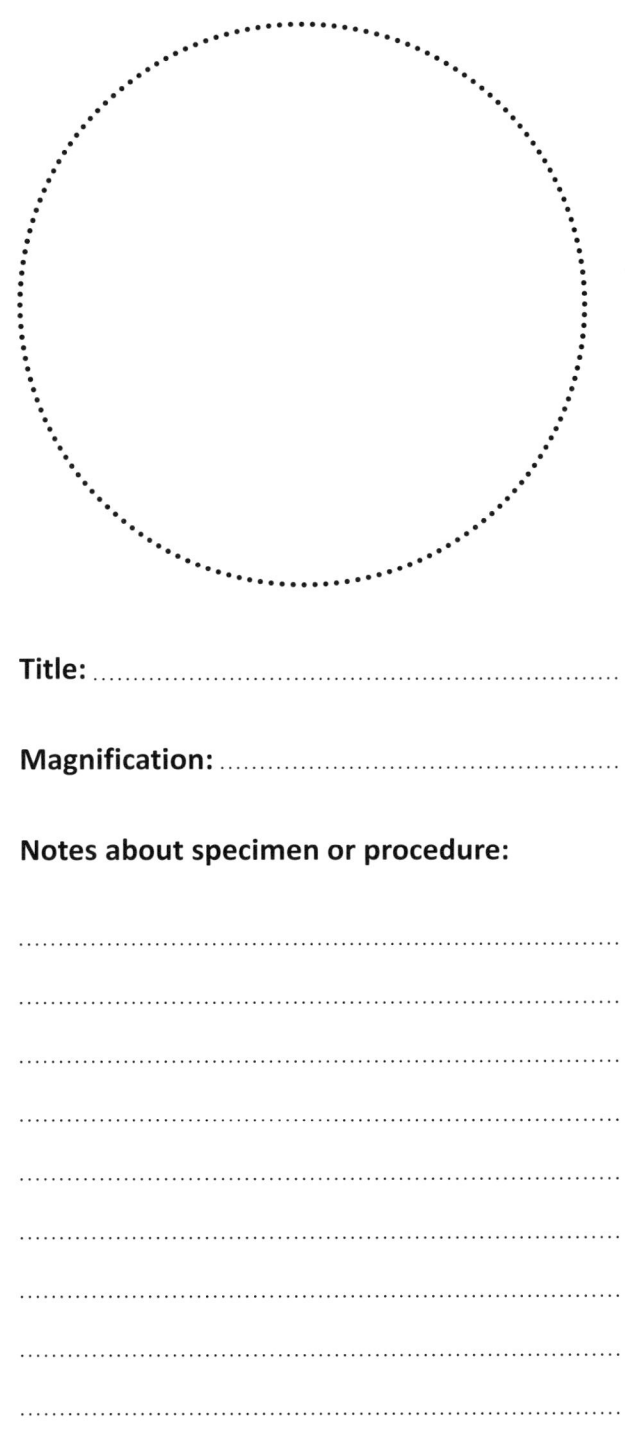

Title: ....

Magnification: ....

Notes about specimen or procedure:

....

....

....

....

....

....

....

....

....

Chapter 17

Workbook 17.1

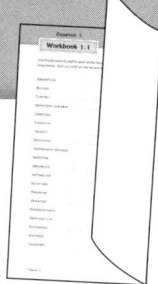

Sample

Use the glossary to define each of the keywords on the lines below. Quiz yourself on the keywords every day until you can explain the meaning of each term without looking at the definition.

- ADIPOSE CELLS ....
- BONE ....
- CARDIAC MUSCLE ....
- CARTILAGE ....
- INVOLUNTARY ....
- JOINT ....
- LIGAMENT ....
- MARROW ....
- PERIOSTEUM ....
- SKELETAL MUSCLE ....
- SMOOTH MUSCLE ....
- STRIATIONS ....
- TENDON ....
- VOLUNTARY ....

Workbook 17.2

Write the letter "V" or "I" on the lines below, depending on whether the action depends mainly on voluntary (V) or involuntary (I) muscles.

1. .... Heartbeat

2. .... Raising your hand

3. .... Hiccups

4. .... Crossing the street

5. .... Opening the door

6. .... Intestinal contractions

7. .... Tapping your finger

8. .... Digesting food

9. .... Riding a bike

Workbook 17.3

Complete each sentence using one of the keywords to fill in the blank.

1. The ball and socket .... allows the most movement.

2. The .... in the ear allows it to bend without breaking.

3. Both skeletal and cardiac muscles have ....

4. Smooth muscle is .... , contracting without conscious thought.

5. An overweight individual has extra ....

6. Branching cells with striations and one nucleus are found in ....

7. The upper and lower leg bones are connected by ....

8. The rubber-band-like material you can feel behind your knee is a .... that connects the thigh muscle to the shin bone.

9. Red and white blood cells are produced in the ....

10. Calcium salts cause .... cells to be hard and rigid.

Workbook 17.4

A. Label the parts of the skeleton indicated below.

1. ....
2. ....
3. ....
4. ....
5. ....

B. List five functions that bones carry out for the body.

....
....
....
....
....
....
....
....

Chapter 17

Workbook 17.5

Diagram, label, and color a knee joint in the space below. Refer to Figure 17.3 in the text. You may use a pencil to draw the diagram, but use a pen to write the labels. Use a ruler for any straight lines, and don't forget to include a title.

Title: ....

Workbook 17.6

Diagram, label, and color three types of muscle cells. Refer to Figure 17.7 in the text. You may use a pencil to draw the diagram, but use a pen to write the labels. Use a ruler for any straight lines, and don't forget to include titles.

Title: ....

Title: ....

Title: ....

Workbook 17.7

"Rubberize" a Bone

Supplies:
- cooked chicken bone
- jar or glass
- white distilled vinegar

1. Clean the chicken bone, being careful to remove any pieces of meat that might be stuck to it. Wash the bone thoroughly with soap and warm water.

2. Put the clean chicken bone in the jar and fill the jar with vinegar.

3. Set the jar aside until the bone is no longer rigid. (If you are using a drumstick, this will probably take seven to nine days.) Remove the bone and rinse off the vinegar. The acetic acid in the vinegar has removed the calcium that was stored in the bone, so the bone will now bend easily!

What just happened?

Calcium is an important component of bones, giving them strength and density. Calcium is found in bones in the form of calcium phosphate. The acetic acid in the vinegar breaks the bonds between the calcium and the phosphorus compound, leaving the calcium free to dissolve. Since white distilled vinegar is mostly water, the calcium simply dissolves out of the bone and into the vinegar. The rubber-like substance that is left is mostly collagen and elastin proteins, which are the main components of cartilage.

Chapter 18
Workbook 18.1

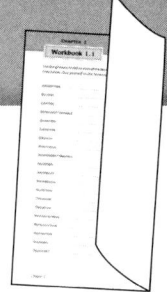

Fold

Sample

Use the glossary to define each of the keywords on the lines below. Quiz yourself on the keywords every day until you can explain the meaning of each term without looking at the definition.

Amylase ....

Anus ....

Bicuspid ....

Bile ....

Chemical Digestion ....

Cuspid ....

Esophagus ....

Gall Bladder ....

Incisor ....

Lacteal ....

Large Intestine ....

Liver ....

Mechanical Digestion ....

Molar ....

Pancreas ....

Pepsin ....

Peristalsis ....

Pyloric Sphincter ....

Rectum ....

Saliva ....

Small Intestine ....

Fold

Chapter 18

Workbook 18.1

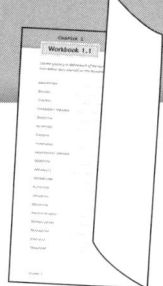

Sample

Use the glossary to define each of the keywords on the lines below. Quiz yourself on the keywords every day until you can explain the meaning of each term without looking at the definition.

STOMACH ....

TASTE BUDS ....

TONGUE ....

VILLUS ....

Workbook 18.2

A. Using complete sentences, explain the difference between mechanical and chemical digestion.

....
....
....
....
....
....
....
....

B. List six different enzymes and tell where each is made and what kind of food it helps to break down.

ENZYME	WHERE IT IS MADE	FOOD IT HELPS BREAK DOWN
....	....	....
....	....	....
....	....	....
....	....	....
....	....	....
....	....	....

Workbook 18.3

Diagram, label, and color the parts of the human digestive system. Refer to Figure 18.9 in the text. You may use a pencil to draw the diagram, but use a pen to write the labels. Use a ruler for any straight lines, and don't forget to include a title.

Title: ....

Workbook 18.4

Use a keyword from this chapter to fill in each blank.

1. A ………………………………, like a root hair, increases surface area for absorption.

2. The ……………………………… is sometimes called a canine tooth.

3. ……………………………… is the enzyme that initiates the chemical digestion of starches in the mouth.

4. Hydrochloric acid must be present for ……………………………… to begin the chemical digestion of proteins in the stomach.

5. Chemical digestion of fats begins after ……………………………… emulsifies them in the small intestine.

6. Food is moved through the digestive system by rhythmic motions called

   ……………………………… .

7. A ……………………………… inside each villus absorbs the fatty acids and glycerol from fat digestion.

8. The ………………………………, the largest organ in the body, makes bile.

9. The ……………………………… stores bile.

10. The ……………………………… is a tube of smooth muscle tissue that connects the mouth to the stomach.

11. Food cannot move out of the stomach until the ……………………………… opens up.

12. Your ……………………………… is needed to move food into the pharynx.

13. Water is reabsorbed inside the ……………………………… .

14. Amylase, trypsin, and lipase are made in the ……………………………… and added to food in the intestine.

15. The stool is formed in the ……………………………… before elimination through the anus.

Workbook 18.5

A. *The following chart summarizes the process of digestion for carbohydrates, fats, and proteins. Fill in the blanks to complete the chart. The first box has been filled in for you.*

SUMMARY OF DIGESTION

	Carbohydrate	Fat	Protein
Where chemical digestion begins	1. mouth	4.	7. ....
Enzyme(s) involved	2.	5.	8. ....
Final product(s)	3.	6.	9. ....

B. *List the four main kinds of teeth, and their functions. Be able to locate them on a diagram.*

1. ....

2. ....

3. ....

4. ....

UPPER JAW

192   Chapter 18

Workbook 18.6

Use a complete sentence to answer each question.

1. What is the name of the thick muscle in the mouth?
   ....
   ....

2. What are the bumps on the muscle in question #1 called?
   ....
   ....

3. What are the four basic tastes?
   ....
   ....

4. How many teeth does an adult human have?
   ....
   ....

5. What is the longest and most important organ of the digestive system?
   ....
   ....
   ....

6. For what are digested foods used by the body?
   ....
   ....
   ....

7. What are the wave-like movements of the digestive system called?
   ....
   ....
   ....

8. What structures are inside villi?

   ....

   ....

   ....

9. How long does food remain in the stomach?

   ....

   ....

   ....

10. What is the name of the valve (ring-like muscle) between the stomach and the small intestine? What is its function?

    ....

    ....

    ....

11. Name two chemicals made by bacteria in the large intestine that are used by the body.

    ....

    ....

    ....

18.7 Formal Lab #5

Complete the experiment below, following the instructions provided. Fill in the blanks as you go. Use complete sentences to answer the questions at the end.

Name: .... Date: ....

I. Title: Chemical Digestion of Starch

II. Purpose: On pages 161-162, the text explains that .... , a protein in our saliva, breaks down starch molecules into sugar molecules. In this experiment we will test this statement. We will test 1) whether starch molecules are really broken down into sugar molecules in the mouth, and 2) whether the larger starch molecules are changed into smaller sugar molecules by the chemical action of saliva or by the mechanical, grinding action of the teeth.

My hypothesis is that starch molecules .... **[are/are not] broken down into sugar molecules in the mouth by** .... **[saliva/teeth].**

III. Materials: The materials required for this experiment are cooked rice, Benedict's solution, and water.

IV. Apparatus: The equipment required for this experiment is at least one glass test tube, a saucepan, a knife, and a spoon.

V. Procedure:

1. Boil 4 inches of water in the saucepan. Turn the heat down to a simmer.
2. Carefully drop a few grains of cooked rice into the bottom of the test tube. Cover the rice with Benedict's solution and heat the test tube for about four minutes by holding the test tube in the pot of simmering water. You may need to use a test tube holder or potholder to hold the test tube. Do not allow the test tube to touch the bottom of the pot.
3. After about four minutes, make a note of the final color of the mixture in Figure 1. Allow the test tube and its contents to cool, then wash and dry it thoroughly.
4. Using a spoon, collect some of your saliva and place it in the test tube. Cover the saliva with Benedict's solution and heat the test tube for about four minutes in the pot of simmering water. The mixture in the test tube should remain blue, unless you have recently eaten something sweet or starchy. Make note of the final color of the mixture in Figure 1. Allow the test tube and its contents to cool, then wash and dry it thoroughly.
5. Place a spoonful of rice in your mouth and chew it for about 30 seconds. Using a spoon, place some of the chewed-up rice in the test tube. Cover the mixture with Benedict's solution and heat the test tube for about four minutes in the pot of simmering water. Make note of the final color of the mixture in Figure 1. Allow the test tube and its contents to cool, then wash and dry it thoroughly.

6. Using a knife, chop up several grains of rice into a paste. Place the rice in the test tube, cover with Benedict's solution and heat the test tube for about four minutes in the pot of simmering water. Make note of the final color of the mixture in Figure 1. Allow the test tube and its contents to cool, then wash and dry it thoroughly.

7. Refer to Figure 15.6 on page 141 in the textbook to fill in the second column of Figure 1.

VI. Data:

	Color after Heating	Amount of Simple Sugar
rice		
saliva		
chewed rice		
chopped rice		

FIGURE 1. RESULTS OF TESTS FOR SIMPLE SUGARS

VII. Questions:

1. Are starch molecules broken down into sugar molecules in the mouth? If so, does the saliva convert the starch molecules into sugar molecules or do the teeth break the large starch molecules into the smaller sugar molecules? Be sure to explain how the experiment supports your answers.

   ....
   ....
   ....
   ....

2. Steps 1 and 2, in which we tested saliva and rice individually for simple sugars, were our controls in this experiment. Why were these controls important to the experiment? (What did we learn about rice and saliva by performing steps 1 and 2?)

   ....
   ....
   ....
   ....

VIII. Conclusion: This experiment showed that starch molecules .... [are/are not] broken down into sugar molecules in the mouth by .... [saliva/teeth].

Thus, my hypothesis was .... [correct/incorrect].

Chapter 19

Workbook 19.1

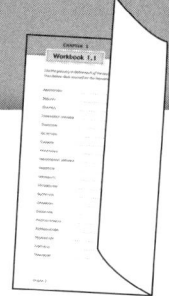

Sample

Use the glossary to define each of the keywords on the lines below. Quiz yourself on the keywords every day until you can explain the meaning of each term without looking at the definition.

Aorta ....

Artery ....

Atrium ....

Capillary ....

Fibrinogen ....

Heart ....

Hemoglobin ....

Inferior Vena Cava ....

Plasma ....

Platelet ....

Pulmonary Artery ....

Pulmonary Circulation ....

Pulmonary Vein ....

Red Blood Cell ....

Septum ....

Superior Vena Cava ....

Systemic Circulation ....

Vein ....

Ventricle ....

White Blood Cell ....

Workbook 19.2

Diagram, label, and color the heart. Refer to Figure 19.3 in the text. You may use a pencil to draw the diagram, but use a pen to write the labels. Use a ruler for any straight lines, and don't forget to include a title.

Title: ....

Extra Credit: If you have not dissected a heart before, obtain a turkey or calf heart from your local butcher or supermarket and dissect it. Compare the heart to your diagram of the human heart and identify the left and right atria, the left and right ventricles, the septum, the aorta, and the veins.

Chapter 19

Workbook 19.3

If a statement below is false, change the italicized word to make it true. Neatly write out each statement once it is true.

1. Waste material from a cell diffuses through the lymph fluid into *capillaries*.

2. Oxygen (O_2) is carried by hemoglobin in the *platelets*.

3. In systemic circulation, oxygen-poor blood returns to the heart through *arteries*.

4. The *veins* have valves.

5. Flowing through the inferior vena cava, the blood enters the right *ventricle*.

6. The right *ventricle* pumps blood to the lungs.

7. Pulmonary *veins* contain oxygen-rich blood.

8. The left *ventricle* is the strongest chamber of the heart.

9. The large artery that curves out of the top of the heart to carry blood to the body is the *superior vena cava*.

   ....
   ....

10. Each artery is *thick*-walled and muscular.

    ....
    ....

11. *Platelet* is the liquid part of the blood.

    ....
    ....

12. *Red* blood cells fight disease organisms.

    ....
    ....

13. *Hemoglobin* is a blood protein needed for clotting.

    ....
    ....

14. Systemic circulation takes blood to the *lungs.*

    ....
    ....

15. Cardiac muscle is *smooth* and has one nucleus in each cell.

    ....
    ....

Workbook 19.4

Complete the chart below.

THE CIRCULATORY SYSTEM

PART	DESCRIPTION	FUNCTION
1.	2.	To dissolve and carry enzymes, gases, and salts
3.	Colorless fragments of cells	4. ....
White blood cells	White amoeboid cells	5. ....
6.	Disk-shaped with no nuclei	7. ....
8.	9.	To transport the blood back to the heart
10.	Thick-walled tube with no valves	11. ....
12.	Muscular, about fist-size	13. ....
14.	Thin-walled, very tiny vessels	To allow diffusion of materials into and out of the blood

CHECK IT OUT! http://www.texasheart.org/ProjectHeart/upload/ph_play_experiments_capillaries.pdf

Workbook 19.5

A. Compose a short paragraph explaining systemic circulation, tracing the path of the blood through the blood vessels and parts of the heart. Be sure to include an explanation of the process of gas exchange between the blood and body cells.

....
....
....
....
....
....
....
....
....
....

B. Write a short paragraph explaining pulmonary circulation, tracing the path of the blood through the blood vessels and parts of the heart. Be sure to include an explanation of the process of gas exchange that occurs within the lungs.

....
....
....
....
....
....
....
....
....
....

Chapter 19

Workbook 19.6

Microscope: Blood Cells

Supplies:
- microscope
- microscope slide and cover slip
- drop of blood

Note: If your microscope came with prepared slides, you may view the prepared blood cell slide instead of preparing your own slide.

1. Place a drop of blood on a microscope slide. Smear the drop by using the cover slip to draw the drop across the slide before lowering the cover slip over the blood.

2. Follow the instructions on page 25 to view the prepared slide through the microscope. If you look at it right away, before the blood dries, you should be able to see the blood cells flowing around each other.

3. Draw what you see at high power in the space below. Be sure to label your drawing and record the magnification of the microscope.

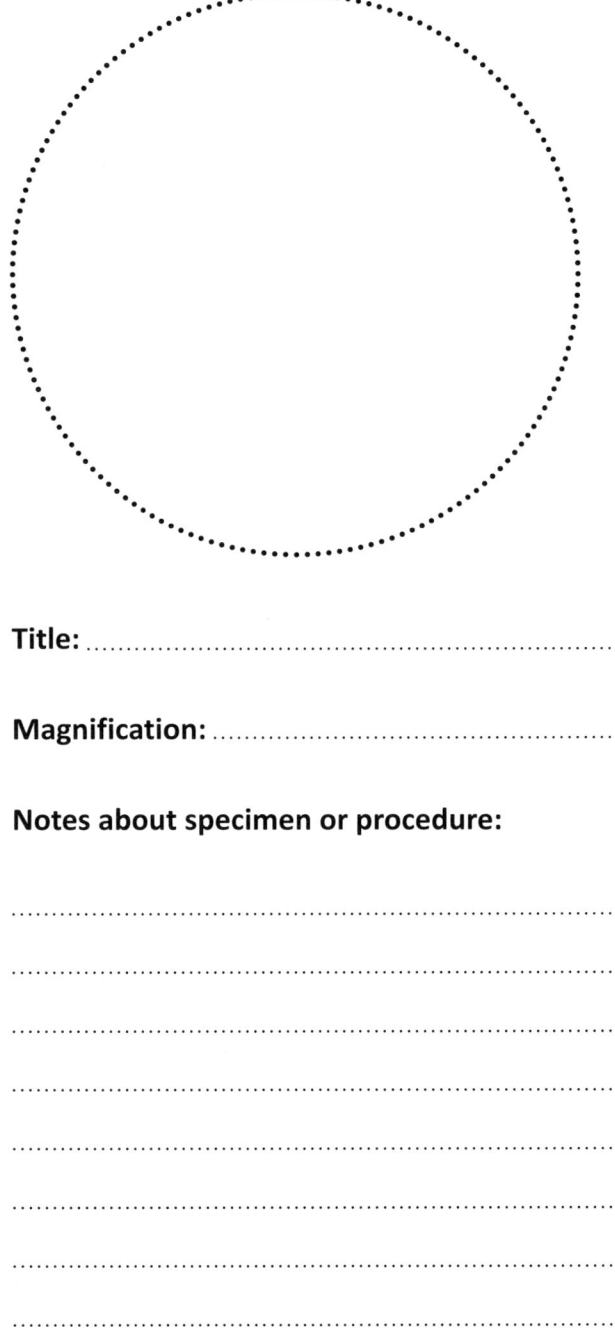

Title: ....

Magnification: ....

Notes about specimen or procedure:

....
....
....
....
....
....
....
....
....
....

Chapter 20

Workbook 20.1

Use the glossary to define each of the keywords on the lines below. Quiz yourself on the keywords every day until you can explain the meaning of each term without looking at the definition.

Don't forget to review your keywords from Chapters 1-19! Frequent review of keywords from previous chapters is critical to doing well on the Final Review.

Alveolus ....

Bronchial Tubes ....

Bronchus ....

Diaphragm (1) ....

Epiglottis ....

Exhale ....

Expiration ....

Hiccups ....

Inhale ....

Inspiration ....

Larynx ....

Mucus ....

Nostrils ....

Oxidation ....

Pharynx ....

Trachea ....

Vocal Cords ....

Workbook 20.2

Diagram, label, and color the respiratory system. Refer to Figure 20.6 in the text. You may use a pencil to draw the diagram, but use a pen to write the labels. Use a ruler for any straight lines, and don't forget to include a title.

Title: ....

Workbook 20.3

Answer the following questions with a brief, but complete, sentence.

1. List the functions of the respiratory system.

   ....

   ....

   ....

2. What gas is required by every living cell in the human body?

   ....

   ....

   ....

3. What gas must be eliminated from every cell?

   ....

   ....

   ....

4. Describe an ideal respiratory surface.

   ....

   ....

   ....

5. What organ do fish use for respiration?

   ....

   ....

   ....

6. What is the respiratory surface in worms?

   ....

   ....

   ....

7. What part of an amoeba is used for respiration?

   ....
   ....
   ....

8. Describe alveoli including what is inside and what is outside of each air sac.

   ....
   ....
   ....

9. What is external respiration?

   ....
   ....
   ....

10. What is cellular respiration?

    ....
    ....
    ....

Workbook 20.4

Use a keyword to fill in each blank.

1. Inside the larynx are two ....

2. When you breathe through the ...., the air is cleaned and warmed.

3. To inhale, the .... and rib muscles must contract.

4. The tar and smoke from cigarettes clog up each .... at the end of the bronchioles so that breathing becomes difficult.

5. Each lung is connected to the trachea by a ....

6. .... is sticky and moist in order to trap dust and germs that may enter the respiratory system.

7. During ...., the diaphragm relaxes and the rib cage is lowered to force air out of the lungs.

8. Even unborn babies can get the .... when their diaphragm spasms.

9. The .... prevents food from entering the trachea.

10. Oxygen is needed in each cell for .... During this process, the oxygen is combined with glucose to release energy.

Workbook 20.5

A. Label the following ten statements in the correct order. The first and last statements have been labeled for you. You may wish to refer to the previous chapter.

1. .... Oxygen is used in the cell.

2. .... Oxygen diffuses into the blood.

3. .... 1 .... Oxygen enters the lungs.

4. .... The blood leaves the right ventricle.

5. .... The blood leaves the left ventricle.

6. .... 10 ...... Carbon dioxide is exhaled from the lungs.

7. .... "Blue" oxygen-poor blood flows in arteries.

8. .... Red oxygen-rich blood flows in the arteries.

9. .... Hemoglobin holds onto oxygen.

10. .... Oxygen diffuses out of the blood.

B. Compose a short paragraph discussing how the skeletal, muscular, circulatory and respiratory systems are interrelated.

....
....
....
....
....
....
....
....
....

Chapter 20

Workbook 20.6

If false, change the italicized word to make the statement true. Neatly write each full statement once it is true.

1. The function of the human *circulatory* system is to exchange gases.

2. The nose and mouth *dry* the air before it reaches the lungs.

3. The *larynx* is the space at the back of the nose and mouth.

4. The epiglottis prevents food from entering the *trachea.*

5. Vibrations of the *bronchial tubes* cause sounds for speech.

6. The *esophagus* directly connects the pharynx to the bronchi.

7. The *trachea* has rings of cartilage to keep it open.

8. The bronchioles end in *alveoli.*

9. Like root hairs and villi, the *bronchi* greatly increase the surface area of an organ.

   ....
   ....

10. When too much carbon dioxide is in the blood, the brain tells the *hiccups* and rib muscles to contract.

    ....
    ....

11. Pulmonary circulation brings blood to the *lungs.*

    ....
    ....

12. The process of oxygen moving from the blood into the cells for use in oxidation is called *cellular respiration.*

    ....
    ....

13. Waste carbon dioxide moves from the cells to the lymph fluid and into the blood by *absorption.*

    ....
    ....

14. The veins and diaphragm are made of *cardiac* muscle.

    ....
    ....

15. *Digestion* occurs when the rib muscles and the diaphragm relax.

    ....
    ....

CHAPTER 21

Workbook 21.1

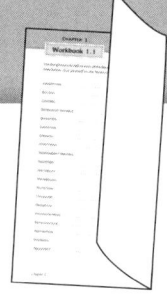

Sample

Use the glossary to define each of the keywords on the lines below. Quiz yourself on the keywords every day until you can explain the meaning of each term without looking at the definition.

DERMIS ......

DIALYSIS ......

EPIDERMIS ......

GLOMERULUS ......

NEPHRON TUBULE ......

NEPHRONS ......

STOOL ......

SWEAT GLANDS ......

UREA ......

URETERS ......

URETHRA ......

URINARY BLADDER ......

URINARY TRACT ......

URINE ......

Fold

Fold

Workbook 21.2

Complete the chart below.

1. .... **SYSTEM**

Part of the Excretory System	Wastes that Are Excreted	Opening(s) for Elimination
2.	CO_2 and H_2O	3. .... 4. ....
Large intestine	5.	6. ....
7.	8. .... 9.	capillaries bile duct
10.	11. .... 12. .... 13.	urethra
14.	heat, water, salts, and nitrogenous wastes	15. ....

Chapter 21

Workbook 21.3

Diagram, label, and color the kidney. Refer to the cross-section of a kidney in Figure 21.3 in the text. You may use a pencil to draw the diagram, but use a pen to write the labels. Use a ruler for any straight lines, and don't forget to include a title.

Title: ....

Workbook 21.4

Fill in the blank using a keyword from this chapter.

1. Inside the kidney are millions of .... that filter the blood.

2. .... occurs naturally in healthy kidneys.

3. Water is reabsorbed into the capillaries that wrap around the ....

4. Perspiration is formed in the .... and excreted through pores in the skin.

5. The sweat glands are in the ....

6. The pores are in the ....

7. Testing a patient's .... can often indicate if the person is healthy.

8. The colorless nitrogen compound .... is formed in the liver from ammonia.

9. The .... connect the kidneys to the urinary bladder.

Workbook 21.5

Write a concise paragraph explaining the excretion of carbon dioxide. Include:

how CO_2 is made inside the body;

where CO_2 is made;

how CO_2 gets to the organ that excretes it;

the organ of excretion for CO_2;

the openings through which CO_2 leaves the body;

what happens if the CO_2 is not eliminated from the body.

Workbook 21.6

Diagram, label, and color a cross-section of the skin. Refer to Figure 21.6 in the text. You may use a pencil to draw the diagram, but use a pen to write the labels. Use a ruler for any straight lines, and don't forget to include a title.

Title: ....

Workbook 21.7

Microscope: Hair

Supplies:
- microscope
- microscope slide and cover slip
- different types of hair (straight, curly, brown, blonde, thin, thick, etc.)

1. Place one strand of hair on a slide and cover it with a cover slip. Examine the slide under the microscope. Be sure to examine the follicle.

2. Repeat using different types of hair. Compare a hair with split ends with one that has been recently trimmed.

3. Draw what you see at high power in the space below. Be sure to label your drawing and record the magnification of the microscope.

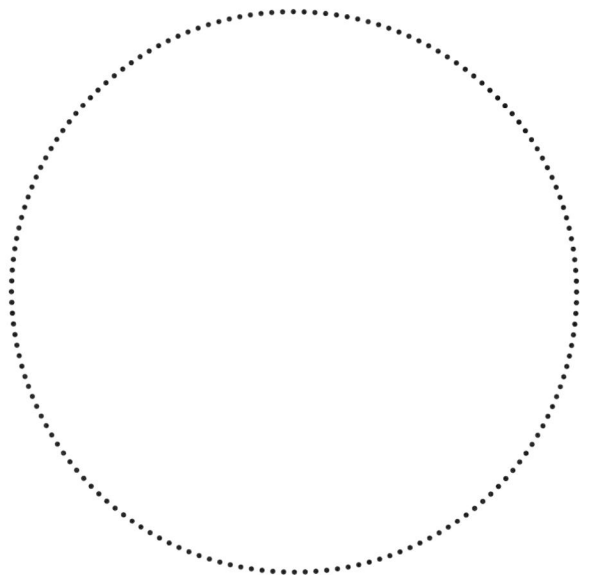

Title: ....

Magnification: ....

Notes about specimen or procedure:

....
....
....
....
....
....
....
....
....
....

Chapter 22

Workbook 22.1

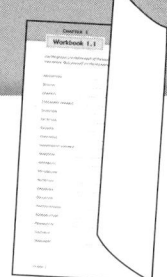

Sample

Use the glossary to define each of the keywords on the lines below. Quiz yourself on the keywords every day until you can explain the meaning of each term without looking at the definition.

- ADRENAL GLANDS ....
- ADRENALIN ....
- CORTISOL ....
- DIABETES ....
- ENDOCRINE SYSTEM ....
- GLUCAGON ....
- GROWTH HORMONE ....
- HORMONES ....
- INSULIN ....
- ISLANDS OF LANGERHANS ....
- PITUITARY GLAND ....
- SIMPLE GOITER ....
- THYROID GLAND ....
- THYROXIN ....

Workbook 22.2

A. List two glands that are not endocrine glands.

....

....

....

....

....

B. In a complete sentence, explain the most important difference between endocrine glands and other glands of the body.

....

....

....

....

....

Workbook 22.3

In the blank spaces, write the name of the gland most associated with the phrase.

pituitary gland   thyroid gland   pancreas   gonads

adrenal gland   parathyroid glands   thymus gland

1. .... Often called the "master gland"

2. .... Butterfly-shaped gland located in the neck

3. .... Requires the mineral iodine in order to work properly

4. .... Secretes thyroxin

5. .... Secretes growth hormone

6. .... Controls the rate at which the body uses calcium

7. .... Diabetes can result if this gland doesn't work properly.

8. .... Secretes cortisol

9. .... Develops and stimulates the immune system

10. .... Located at the base of the brain

11. .... Located on top of the kidneys

12. .... The sex glands (ovaries and testes)

13. .... Secretes a hormone that causes the liver to convert glycogen back into glucose and release it into the blood

14. .... Both a digestive gland and an endocrine gland

15. .... Often called the "glands of combat"

16. .... Goiter and cretinism are two conditions that can result when this gland does not work properly.

Workbook 22.4

List five endocrine glands and indicate their location in the body.

Name of Endocrine Gland	Location in the Body
....	....
....	....
....	....
....	....
....	....

Workbook 22.5

Make a chart listing five diseases related to the endocrine system. Include the cause of the disease and the hormone involved.

Disease	Cause	Hormone

Workbook 22.6

Research on a Dwarf or a Giant

Choose a dwarf or a giant to learn more about. Write a paragraph and share your findings with your family. Here are a few for you to choose from:

Dwarfs

- Charles Stratton (Tom Thumb)
- Pauline Musters
- Calvin Phillips
- Michu
- William Phillips
- William E. Jackson
- Caroline Crochami
- Max Taborshy

Giants

- Robert Wadlow
- K. A. Jabbar
- Ivan Lushkin
- Sandy Allen
- Jane Bunfor

Chapter 23

Workbook 23.1

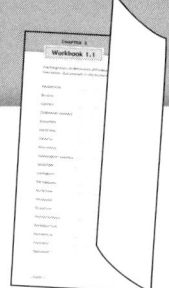
Sample

Use the glossary to define each of the keywords on the lines below. Quiz yourself on the keywords every day until you can explain the meaning of each term without looking at the definition.

- Associative Neuron ....
- Auditory Nerve ....
- Axon ....
- Cerebellum ....
- Cerebrum ....
- Cochlea ....
- Cone Cells ....
- Convolutions ....
- Cornea ....
- Cyton ....
- Dendrites ....
- Effectors ....
- Eustachian Tube ....
- Farsightedness ....
- Impulse ....
- Iris ....
- Medulla ....
- Motor Neuron ....
- Nearsightedness ....
- Neuron ....
- Neurotransmitter ....

Chapter 23

Workbook 23.1

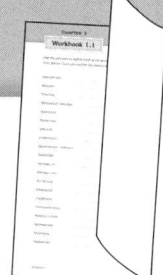
Sample

Use the glossary to define each of the keywords on the lines below. Quiz yourself on the keywords every day until you can explain the meaning of each term without looking at the definition.

OLFACTORY NERVE ....

OPTIC NERVE ....

RECEPTORS ....

RETINA ....

ROD CELLS ....

SEMICIRCULAR CANALS ....

SENSORY NEURON ....

SYNAPSE ....

Workbook 23.2

List the eight functions of the nervous system and the part of the body involved.

Name of Function	Part of the Body
....	....
....	....
....	....
....	....
....	....
....	....
....	....
....	....

Workbook 23.3

Diagram, label, and color a neuron. Refer to Figure 23.3 in the text. You may use a pencil to draw the diagram, but use a pen to write the labels. Use a ruler for any straight lines, and don't forget to include a title.

Title: ....

Workbook 23.4

Fill in the blanks with the correct type of neuron.

1. .... neurons bring messages from the brain to a muscle.

2. .... neurons are concentrated in the sense organs.

3. .... neurons are found mostly in the brain.

4. .... neurons transmit messages from the eye to the brain.

5. .... neurons carry the impulses that cause a gland to secrete a hormone.

6. .... neurons are located between the other two kinds of neurons.

7. .... neurons have specialized parts called receptors.

8. .... neurons are connected to effectors.

9. .... neurons form ideas.

10. Paralysis results when the .... neurons are cut.

Workbook 23.5

A. List the three main parts of the human brain and the functions each part controls.

Part of the Brain	Functions
....	....
....	....
....	....

B. List the five sense organs, the name of the sense, and the special nerve(s) associated with each sense. If the nerve is not named in the chapter, list the part(s) containing the nerve.

Sense Organ	Name of the Sense	Special Nerves or Parts
....	....	....
....	....	....
....	....	....
....	....	....
....	....	....

Workbook 23.6

Label the parts of the eye.

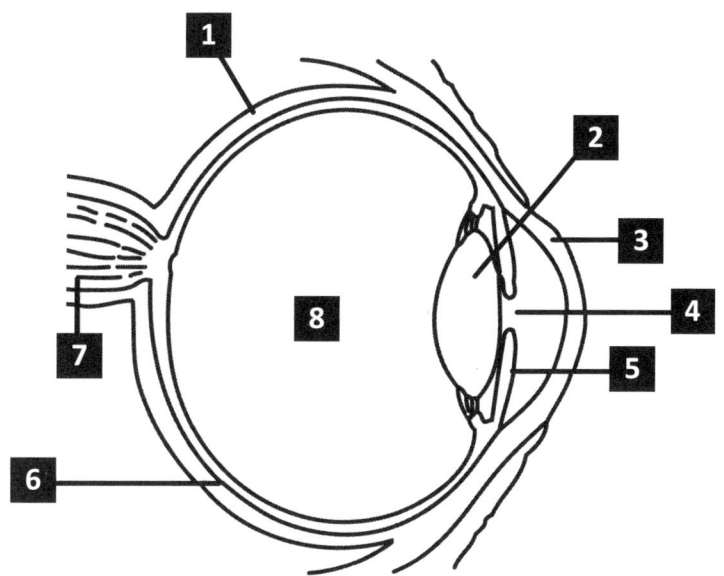

1. ....
2. ....
3. ....
4. ....
5. ....
6. ....
7. ....
8. ....

CHECK IT OUT! Cow's eye dissection: *http://www.exploratorium.edu/learning_studio/cow_eye/*

Workbook 23.7

Write a short paragraph distinguishing between nearsightedness and farsightedness. Be sure to use at least six of the following terms:

concave lens	farsightedness	receptors
cone cells	iris	retina
convex lens	nearsightedness	rod cells
cornea	optic nerve	sensory neuron

....

....

....

....

....

....

....

....

....

Workbook 23.8

Use a keyword from Chapter 23 to fill in the blanks.

1. The .... connects the eye to the brain.

2. The rod cells and cone cells together make up the ....

3. .... are the branch-shaped parts of a neuron.

4. The .... is the snail-shaped part of the ear.

5. People who are color blind may have a problem with their ....

6. The number of ...., bumps, on the cerebrum may be connected to how smart a person is.

7. The nerve message crosses the .... by means of a neurotransmitter.

8. Damage to the brain's .... could cause a loss of balance and coordination.

9. Thinking out the answers to these questions occurs in your ....

10. The .... is also called the brain stem.

11. Air pressure behind the eardrum is changed when air moves through the ....

12. .... is a condition in which the eye focuses the light too far from the lens.

13. A person's eye color is determined by the color of his ....

14. The aroma of a hot apple pie causes the .... to send impulses to the brain.

15. The fluid in the .... moves when you move and impulses are sent to the cerebellum.

16. Like the terms "auditorium," "audition," and "audio," the .... has to do with hearing.

17. The .... of a neuron is enclosed in a fatty sheath and can be as long as 1.5 m.

Chapter 23 239

23.9 Formal Lab #6

Complete the experiment below, following the instructions provided. Fill in the blanks as you go. Use complete sentences to answer the questions at the end.

Name: .... Date: ....

I. Title: Distance between Nerves

II. Purpose: The skin contains many nerves with touch receptors. Surfaces of the body that need to be more sensitive than others have a greater number of nerves, which are placed closer together than in other parts of the body.

A single nerve can only send one signal to the brain at a time. Thus, even if two things are touching it, it sends the same signal as it would when one thing is touching it. We can use this information to discover whether the nerves are closest together in the palm, shoulder, or calf.

My hypothesis is that the nerves in the .... **[palm/shoulder/calf] are placed closest together.**

III. Materials: The materials required for this experiment are two paperclips and a partner.

IV. Apparatus: The equipment required for this experiment is a centimeter ruler.

V. Procedure:

1. Unbend the two paperclips into the shapes shown in the figure below.
2. Adjust the ends of paperclip B so that the ends are 4 cm apart. Use a centimeter ruler to measure the distance between the ends.
3. Tell your partner that you are going to touch different parts of his body with two paperclips: paperclip A, which has one point, and paperclip B, which has two points. Explain that his job will be to identify, without looking, which paperclip is which.
4. Ask your partner to close his eyes and hold out his hand, palm facing up. Touch his palm with the two ends of paperclip B. Then touch his palm with the single end of paperclip A. Ask him to

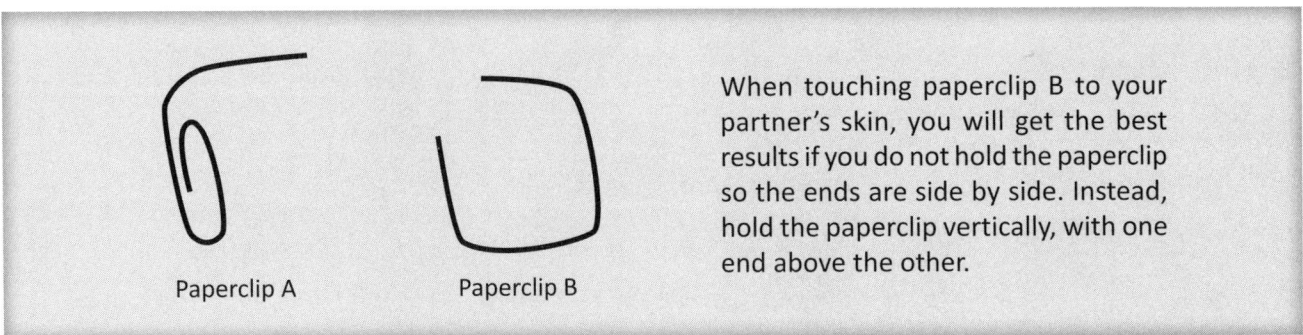

tell you whether the first or the second paperclip was the one with two points. If he answers correctly, it means that he felt both ends of paperclip B separately. In this case, you should write a "2" in the appropriate space in Figure 1, because he felt *two* points. If he is not able to answer correctly, it means that he felt the two points of paperclip B as a single point, so you should write a "1" in Figure 1.

5. Now repeat step 4 with your partner's shoulder and calf. Be sure to touch his skin with the two paperclips in a different order each time (I.e., don't always start with paperclip B.). For each area of the body, record whether or not he can tell the difference between the two paperclips by writing a "1" or a "2" in Figure 1.

6. Next, adjust paperclip B so that the ends are 3 cm apart instead of 4 cm. Repeat steps 4 and 5.

7. Continue decreasing the distance between the two ends of the first paperclip by 1 cm and repeating steps 3 and 4 until the ends of the first paperclip are nearly touching (about 1-2 mm apart). Remember to record your results by writing a "1" or a "2" in Figure 1. If your partner only feels one paperclip point, it means that both ends of the paperclip are touching a single nerve.

8. Use the data in Figure 1 to fill out Figure 2. For instance, let's say that your partner felt two points on his palm when the paperclip's ends were 2 cm apart, but only felt one point when the paperclip's ends were 1 cm apart. In this case, you know that the distance between the neurons on his palm must be greater than 1 cm, but less than 2 cm. Thus, you would write "1–2 cm" in the first space in Figure 2.

9. Now have your partner perform steps 4-8 on you and record the results in Figures 1 and 2.

VI. Data:

	Palm		Shoulder		Calf	
	Partner	Self	Partner	Self	Partner	Self
4 cm						
3 cm						
2 cm						
1 cm						
1–2 mm						

FIGURE 1. NUMBER OF PAPERCLIP POINTS FELT AT DIFFERENT LOCATIONS ON THE BODY

	Palm	Shoulder	Calf
Distance between neurons	Partner: Self:	Partner: Self:	Partner: Self:

FIGURE 2. DISTANCE BETWEEN NEURONS ON DIFFERENT PARTS OF THE BODY

VII. Questions:

1. The area of the body in which the neurons are closest together has the most neurons. According to the data in Figure 2, in which area of your partner's body are the neurons closest together? In which area of your own body are the neurons closest together?

   ....
   ....
   ....
   ....
   ....
   ....

2. Why do some parts of the body need to be more sensitive—that is, have more neurons—than others?

   ....
   ....
   ....
   ....
   ....

VIII. Conclusion: This experiment showed that the nerves in the .... [palm/shoulder/calf] were placed closest together.

Thus, my hypothesis was .... [correct/incorrect].

Chapter 24

Workbook 24.1

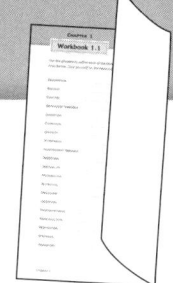

Sample

Use the glossary to define each of the keywords on the lines below. Quiz yourself on the keywords every day until you can explain the meaning of each term without looking at the definition.

AIDS ....

Allergy ....

Antibiotic ....

Antibodies ....

Antiseptic ....

Direct Contact ....

Disease ....

Hemophilia ....

Host ....

Immunity ....

Indirect Contact ....

Infectious ....

Malaria ....

Noninfectious ....

Pathogen ....

Phagocytes ....

Puncture ....

Scurvy ....

Tetanus ....

Vaccine ....

Vector ....

Workbook 24.2

Compose a short paragraph describing the first vaccination ever administered.

Workbook 24.3

Use complete sentences to answer each of the questions below.

1. What are the two basic types of disease?

   ....
   ....

2. List four kinds of noninfectious diseases and provide an example of each.

   ....
   ....
   ....
   ....

3. What are the five ways that infectious diseases can be spread?

   ....
   ....
   ....
   ....
   ....

4. What are the human body's three lines of defense against disease?

   ....
   ....
   ....

5. List the four functions of antibodies.

   ....
   ....
   ....
   ....

Workbook 24.4

List Koch's Postulates.

Workbook 24.5

Complete each sentence using one of the keywords from Workbook 24.1 and a disease from Workbook 24.6 to fill in the blanks. Words will be used only once.

1. Whenever a person has a .... wound, ... is a possible problem.

2. Bleeder's disease, ..., is .... and cannot become an epidemic.

3. The female <u>Anopheles sp.</u> mosquito is the .... that carries ....

4. In countries without meat inspection, you could become the .... to a .... !

5. ...., acquired immunodeficiency syndrome, is spread by .... with infected body fluids.

6. An .... is when a person's system of .... overreacts.

7. .... is a fatal .... that often results from smoking cigarettes.

8. ...., caused by a lack of vitamin C, cannot be prevented by a .... as can smallpox.

Workbook 24.6

Write an "I" (infectious) or an "N" (noninfectious) in the space in front of each disease to identify the type.

1. .... AIDS

2. .... Allergy

3. .... Beef tapeworm

4. .... Hemophilia

5. .... Lung cancer

6. .... Malaria

7. .... Ringworm

8. .... Scurvy

9. .... Tetanus

Workbook 24.7

Match the disease in Column A with the organism which causes it from Column B. You may use a term from Column B more than once.

Column A

1. .... Tapeworm
2. .... Typhoid fever
3. .... Malaria
4. .... Thrush
5. .... Polio
6. .... Smallpox
7. .... Lyme disease
8. .... Amoebic dysentery
9. .... Athlete's foot
10. .... Poison ivy rash

Column B

A. Virus

B. Bacterium

C. Protist

D. Fungus

E. Plant

F. Flatworm

G. Nematode

Workbook 24.8

Complete the chart below.

SCIENTISTS FIGHT FOR LIFE

Name	Date	Notes on Disease or Discovery
1.	1796	2. ....
3.	1865	First to use an antiseptic to clean his surgical tools
4.	1882	Following a series of steps now named after him, he isolated a particular bacterium and proved that it causes tuberculosis.
5.	1885	6. ....
7.	8.	Developed the first antitoxin for use against diphtheria
9.	1928	10. ....
Charles G. King	1932	11. ....
Gerhard Domagk	12.	13. ....
Jonas Salk	1953	14. ....
Willy Burgdorfer	15.	Identified the spirilli bacterium that causes Lyme disease

Chapter 24

Workbook 24.9

Match the disease in Column A with the phrase from Column B that is most closely associated with the disease. You will not use all of the phrases in Column B.

Column A

1. .... Anemia
2. .... Beriberi
3. .... Color blindness
4. .... Diabetes
5. .... Goiter
6. .... Hemophilia
7. .... Lung cancer
8. .... Malaria
9. .... Night blindness
10. .... Rickets
11. .... Scurvy
12. .... Typhoid fever

Column B

A. vitamin C needed

B. vitamin B_1 from grains needed

C. carotene will help

D. vitamin D in milk is needed

E. often caused by smoking

F. Fe is lacking in the diet

G. glucose is not controlled due to lack of insulin

H. "Hemo-" refers to the blood

I. iodine is lacking in the diet

J. Jenner used a vaccine

K. Koch's postulates

L. Louis Pasteur saved a boy

M. mosquitoes are the vectors

N. poor hygiene by food handlers

O. overactive pituitary in child

P. inherited by boys more than girls

Workbook 24.10

Write a short paragraph explaining the difference between natural and acquired immunity. Distinguish between the two types of acquired immunity: active and passive.

Fight against Disease Research Paper

Choose a scientist who made a contribution to our understanding of disease. Consult with your teacher on the topic and due date for your report. Please also reference "Tips for Writing a Research Paper" on pages 143–146 in this workbook. Be sure to outline your topic before writing your first draft.

The paper will include the following information about your selected scientist:

- when and where the scientist lived

- Describe the major accomplishments of the scientist. Explain how his discoveries improved our understanding and treatment of disease. If the scientist is known for curing a specific illness, include a description of the disease, identifying its type (deficiency disease, infectious disease, etc.) and its cause.

- obstacles the scientist overcame

Length of Text

At least 750 words, on a minimum of three pages (not including Table of Contents, glossary, Works Cited)

Format

Follow directions given in your English text, a text of your choice, or in "Tips for Writing a Research Paper" on pages 143–146.

What else to include in your paper:

- Use a minimum of five vocabulary words from the chapter on disease (Chapter 24) and/or the chapters on human anatomy (Chapters 16–23).

- Include a Works Cited page with a minimum of three different sources. (These sources might include library books, magazine articles, the encyclopedia, and one internet article.)

- Include at least three quotations from your sources. See "Tips for Writing a Research Paper" on pages 143–146 for instructions on formatting quotations.

- Provide a glossary defining the minimum of five vocabulary words that were used in your paper.

Putting It All Together:

Your paper should be assembled in this order:

- **Cover** with artwork related to the topic *(optional)*
- **Title page**
- **Table of Contents** listing the following parts of the paper along with the page number on which the information begins
- **Text** of the paper
- **Glossary**
- **Works Cited** page

Please use the information above as a check list. Consult it as you write your paper to be sure that you have included all required information.

Chapter 25

Workbook 25.1

Use the glossary to define each of the keywords on the lines below. Quiz yourself on the keywords every day until you can explain the meaning of each term without looking at the definition.

Don't forget to review your keywords from Chapters 1-24! Frequent review of keywords from previous chapters is critical to doing well on the Final Review.

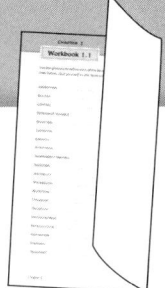

Sample

CONDITIONED REFLEX ....

CONJUGATION ....

DOMINANT ....

EGG ....

GENES ....

GENETICS ....

HYBRID ....

INSTINCT ....

LEARNING ....

MEIOSIS ....

OVARIES ....

PURE ....

REASONING ....

RECESSIVE ....

RESPONSE ....

SPERM ....

STIMULUS ....

TESTES ....

UNCONDITIONED REFLEX ....

Fold

Fold

Workbook 25.2

Identify which type of behavior is most closely associated with each statement or phrase. Write the appropriate letter in front of each statement.

C = Conditioned reflex

I = Instinct

L = Learning

R = Reasoning

U = Unconditioned reflex

1. .... A girl sits on a thorn and quickly jumps up again.

2. .... An oriole builds a hanging, sack-like nest.

3. .... A student washes his hands more often after learning how germs are spread.

4. .... After hitting the canvas from an upper cut, the boxing student no longer leaves his jaw unguarded.

5. .... A frog goes into estivation as its pond dries up in late summer.

6. .... Pavlov's dogs

7. .... Solving a math problem

8. .... Reading a book

9. .... Honey bees building a honey comb

Workbook 25.3

Write a short paragraph explaining how a simple reflex occurs. Include the following terms:

associative neuron	response	axon	unconditioned reflex
sensory neuron	effector	spinal cord	
motor neuron	stimulus	receptor	

Workbook 25.4

Write the following terms in order, beginning with the fertilized egg:

adult   baby   fetus   child   embryo   zygote

....

....

....

....

....

....

Workbook 25.5

Fill in the blanks with the best word or phrase: *meiosis, mitosis,* or *both processes.*

1. The end result of .... is two daughter cells.

2. .... is/are essential for sexual reproduction in many organisms.

3. In ...., the chromosomes are duplicated only once.

4. Cells resulting from .... contain half the usual number of chromosomes.

5. .... result(s) in two cells identical to the parent cell.

6. .... occur(s) for growth.

7. Cells resulting from .... are called gametes.

Workbook 25.6

Fill in the blanks to complete each sentence.

1. Pulling the hand away from a hot pan is an example of an ....

2. A moth builds its cocoon by ....

3. The .... for Pavlov's dogs to begin salivating was a bell.

4. .... from your mistakes is important.

5. Zygomycetes carry out .... as a simple form of sexual reproduction.

6. A .... is any cell with only one half the normal number of chromosomes.

7. The .... cell is produced in the testes.

8. The .... cell is the larger gamete because it contains stored food.

9. A flower with both chromosomes for white blossoms is called ....

10. A person with brown eyes who also has the trait for blue eyes is a ....

11. .... genes always "win out" over recessive genes.

Workbook 25.7

1. Let us assume that the gene for sickle cells (s) is recessive, while the gene for normal blood cells (S) is dominant. A man who has one normal gene (S) and one gene for sickle cells (s) marries a woman with one normal gene (S) and one gene for sickle cells (s). Fill out the Punnett square below to illustrate the probability that their child will have sickle cell anemia.

   Father
	S	s
S		
s		

   Mother

   S = normal   s = sickle cells

2. The ability to taste a substance known as phenylthiocarbamide (PTC) is inherited as a dominant gene (P), while the inability to taste PTC is inherited as a recessive gene (p). If a man is pure for the dominant gene (PP) and a woman is pure for the recessive gene (pp), will their children be able to taste PTC? Draw a Punnett square to illustrate what kind of offspring these parents could have, and then write your answer on the line below.

   ....

3. In peas, yellow seeds (Y) are dominant over green seeds (y). Using the information in the Punnett square below, calculate the probability that the offspring of a plant pure for green seeds (yy) and a plant hybrid for yellow seeds (Yy) will have green seeds.

   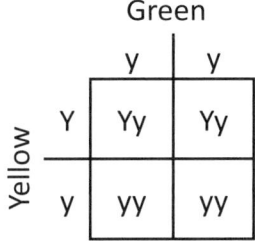

   Green
	y	y
Y	Yy	Yy
y	yy	yy

   Yellow

   Y = yellow seeds   y = green seeds

   Probability of offspring with green seeds = ... %

25.8 Formal Lab #7

Complete the experiment below, following the instructions provided. Fill in the blanks as you go. Use complete sentences to answer the questions at the end.

Name: .... **Date:** ....

I. Title: Sensory Response Times

II. Purpose: Most stimuli are received by the senses of sight, hearing, and touch. For instance, you might see your best friend running towards you, hear him calling your name, and feel your palm stinging when he gives you a "high five." You are able to respond to these stimuli because the sensory neurons in your eyes, ears, and skin send electrical impulses to your brain. In response, your brain sends electrical impulses to motor neurons which cause the muscles in your mouth to smile, the muscles in your legs to turn your body around, and the muscles in your arm to move your hand towards your friend's.

The brain—and in the case of unconditioned reflexes, the spinal cord—can usually respond to stimuli from any of the senses within less than one second, but some responses are faster than others. In this experiment we will attempt to discover whether a person's auditory, visual, or tactile response is fastest.

My hypothesis is that the response time is shortest for the sense of .... **[sight/hearing/touch]**.

III. Materials: No materials are required for this experiment.

IV. Apparatus: The equipment required for this experiment is a centimeter ruler (at least one foot long; a meter stick would be ideal) and four friends or family members.

V. Procedure:

1. Stand facing a friend or family member. Ask your partner to hold his hand in front of him. Hold the ruler in front of you vertically so that the end marked with "0 cm" is at the bottom and hangs between the thumb and fingers of your partner's outstretched hand. Make sure your partner's hand does not touch the ruler.
2. Tell your partner that you are going to drop the ruler, and that he should pinch his fingers and thumb together to catch it as quickly as he can.
3. Drop the ruler so that it falls between your partner's fingers and thumb. When your partner catches the ruler, record in Table 1 how far the ruler fell before he caught it. For instance, if his fingers pinched the ruler after it had fallen 15 cm, you should write "15 cm" in the Visual Response column of Table 1.

4. Next, hold the ruler as before but ask your partner to close his eyes. Instruct him to catch the ruler when he hears you say, "Go!" Drop the ruler while saying, "Go!" (Be sure to drop the ruler and say "Go!" at exactly the same time.) Record the measurement in the Auditory Response column of Table 1.

5. Now hold the ruler so that your partner can gently feel the ruler against his fingers. Tell him to close his eyes and to catch the ruler as soon as he feels it falling. Drop the ruler and record the measurement in Table 1 as before.

6. Repeat steps 1–5 with three more people. Then have someone drop the ruler for you so that you can measure your own response times.

7. Now calculate the averages for each column in Table 1. To do this, add the numbers in the "Visual Response" column together, then divide the sum by five (the total number of people). Record this number in Table 2. Repeat for the "Auditory Response" column and the "Tactile Response" column.

VI. Data:

Name	Visual Response	Auditory Response	Tactile Response

Table 1. Distances the ruler fell during the time it took volunteers to respond to visual, auditory, and tactile stimuli

Visual Response	Auditory Response	Tactile Response

Table 2. Average distance which the ruler fell during the time it took volunteers to respond to visual, auditory, and tactile stimuli

VII. Questions:

1. Compare the average distances in Table 2. Did one sense have a faster response time on average than the others? If so, which one?

....

....

....

....

Chapter 25

2. What are some possible sources of error in this experiment? For instance, is one of the volunteers hard of hearing, or was the experiment performed in a visually distracting location?

....

....

....

....

VIII. Conclusion: This experiment showed that the response time is shortest for the sense of

.... [sight/hearing/touch] .

Thus, my hypothesis was .... **[correct/incorrect].**

Chapter 26

Workbook 26.1

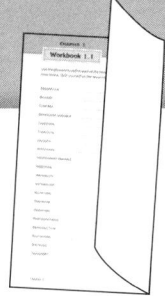

Sample

Use the glossary to define each of the keywords on the lines below. Quiz yourself on the keywords every day until you can explain the meaning of each term without looking at the definition.

ABIOTIC ....

BIOME ....

BIOSPHERE ....

BIOTIC ....

DECOMPOSERS ....

ECOLOGY ....

ECOSYSTEM ....

FOOD CHAIN ....

FOOD WEB ....

HABITAT ....

NICHE ....

PRIMARY CONSUMERS ....

PRODUCERS ....

SECONDARY CONSUMERS ....

Workbook 26.2

A. On the line, label each object as either abiotic or biotic.

1. .... a glass fish tank
2. .... water in the aquarium
3. .... amoebas in the water
4. .... gravel on the bottom
5. .... Elodea, a plant

6. .... a snail
7. .... a catfish
8. .... the filter
9. .... tadpole
10. .... a water beetle

B. Use a keyword to complete each sentence.

1. The energy relationship between a bird, a grasshopper, and grass is an example of a ....

2. A cow's .... includes its role as a primary consumer.

3. N_2 gas is an .... factor in the environment.

4. Every food chain starts with ....

5. Herbivores may be the .... and eat plants.

6. The .... of the haircap moss is any moist shaded area in the forest.

7. All the plants, animals and non-living factors in an area that go together is an ....

8. .... is the study of relationships.

9. The reason there are usually only a few eagles in an area is because eagles are ....

10. Cacti are producers in the desert, one of the six main ....

Chapter 26

Workbook 26.3

Draw a food chain for each set of organisms, using arrows to show the direction of energy flow.

1. cow, grass, man

   grass → cow → man

2. chicken, leafhopper, leaf

   leaf → leafhopper → chicken

3. corn, mouse, snake

   corn → mouse → snake

4. ants, apple, woodpecker

   apple → ants → woodpecker

5. bass, green algae, tadpole

   green algae → tadpole → bass

Workbook 26.4

Use the five food chains from Workbook 26.3 to make one possible food web. Include at least 13 organisms. You do not need to include all possible energy relations between the organisms. Place the producers at the bottom of the web and the secondary consumers at the top of the web. Use arrows to show the flow of energy.

Title: ....

Workbook 26.5

A. Complete this chart of the six biomes using the words below as examples of plants and animals for each biome. The first biome has been done for you.

antelope	bald eagle	boa constrictor	roadrunner	rubber tree
buffalo	sidewinding adder	grizzly bear	beaver	lichen
orchid	saguaro cactus	tumbleweed	pine tree	
prairie grass	Venus flytrap	sphagnum moss	arctic poppy	
caribou	lion	parrot	muskox	

EXAMPLES OF WILDLIFE IN EACH BIOME

Biome	Plant	Animal
Deciduous forest	Live oak	Raccoon
Tundra		
Boreal Forest		
Tropical Rainforest		
Desert		
Grassland		

B. Rearrange the following terms in order from smallest to largest:

biome   habitat   biosphere   ecosystem   organism

....

....

....

....

....

Workbook 26.6

Diagram, label, and color the nitrogen cycle. Refer to Figure 26.6 in the text. You may use a pencil to draw the diagram, but use a pen to write the labels. Use a ruler for any straight lines, and don't forget to include a title.

Title: ....

Final Review

The questions in the Final Review touch on the major topics you have studied in Chapters 14-26. Parts II-IV of the Final Review are designed to be given "closed-book." So that you will be prepared to do your best, review your text carefully, paying special attention to key terms, chemical equations, and charts. You should know the meaning and correct spelling of all the keywords in Chapters 1-26.

Part I. Keywords

A. *Spelling*

1. After you have studied the spelling and definitions of the keywords from Chapters 1-26, ask someone to test you orally on each word using the keyword and definition lists in the workbook. When he reads the definition, correctly spell the keyword on a separate sheet of paper.
2. Restudy any of the keywords you missed and repeat step 1 until you know them all.

B. *Definitions*

1. Now have someone test you on the definitions. When he says the keyword, correctly explain its meaning.
2. Restudy any of the keywords you missed and repeat step 1 until you know them all.

Final Review

Part II. Nutrition and Disease

1. Write the equation for cellular respiration.

2. Write the equation for photosynthesis.

3. What are the five major nutrient groups?

4. List five vitamins or minerals and what part of the body each affects.

5. Explain how to test for the presence of starch in a food sample.

6. Identify four different kinds of noninfectious disease and give an example for each.

Final Review

7. Name five ways infectious diseases may be transmitted.

   ....

   ....

   ....

8. List two diseases caused by each group: viral, bacterial, protistan, fungal and animal (flatworm and nematode).

   ....

   ....

   ....

   ....

9. Identify the major accomplishment(s) of each scientist listed below.

   Jenner ....

   Pasteur ....

   Koch ....

   Von Behring ....

   Reed ....

   Fleming ....

   Domagk ....

   Salk ....

Final Review

Part III. Body Systems

1. List nine systems of the body and the life function that each carries out.

   ....

   ....

   ....

   ....

   ....

   ....

   ....

   ....

   ....

2. List the four types of joints and provide an example of each.

   ....

   ....

   ....

   ....

3. Describe the difference in function between voluntary and involuntary muscles.

   ....

   ....

   ....

4. List two enzymes involved in digestion. Then identify where they are found in the body, and what type of food they help to digest.

   ....

   ....

   ....

   ....

Final Review

5. Identify and describe the four components of blood.

   ....
   ....
   ....
   ....

6. Explain the difference between external respiration and cellular respiration.

   ....
   ....
   ....
   ....

7. On the lines below, identify the wastes excreted by the following excretory organs.

   Lungs: ....
   Large intestine: ....
   Urinary bladder: ....
   Skin: ....

8. List four endocrine glands on the lines below. For each gland, identify a hormone which it secretes and explain the hormone's function.

   ....
   ....
   ....
   ....

9. List the three main parts of the brain. Briefly describe the functions of each part.

   ....
   ....
   ....

Final Review

Part IV. Animal Behavior, Reproduction, and Ecology

1. Explain the difference between a conditioned and an unconditioned reflex.

   ....
   ....
   ....

2. Explain the difference between meiosis and mitosis.

   ....
   ....
   ....
   ....
   ....

3. A female cat is **hybrid** for short hair. A male cat is **pure** for long hair. Using "H" for the dominant short-hair gene and "h" for the long-hair gene, fill in the Punnett square below to calculate the likelihood that the cats will have short-haired or long-haired kittens.

   H = short-haired

   h = long-haired

   % short-haired ....

   % long-haired ....

4. Make a food chain, identifying the producer, primary consumer, and secondary consumer. Also include the producer's energy source and the decomposers.

Final Review

Part V. Diagram of Ear

Diagram, label, and color the ear in the space below. Refer to Figure 23.13 in the text. You may use a pencil to draw the diagram, but use a pen to write the labels. Use a ruler for any straight lines, and don't forget to include a title.

Title: ....

Final Review

Part VI. Diagram of Carbon Cycle

Diagram and label the carbon cycle in the space below. Refer to Figure 26.5 in the text. You may use a pencil to draw the diagram, but use a pen to write the labels. Use a ruler for any straight lines, and don't forget to include a title.

Title: ....

A Miniature Ecosystem

Build your own miniature ecosystem, following the instructions below. A basic, ten-gallon fish tank can be purchased for less than $15; however, the less expensive the fish tank, the greater the chance that it may crack or leak, so be sure to keep the ecosystem in a place where it will cause minimal damage if the tank cracks or leaks. You should never attempt to move a fish tank unless it is completely empty.

A miniature ecosystem is a wholly self-sustaining, biologically balanced collection of natural elements which can be obtained from the nearest stream or pond. All you need is a net (mine was homemade) and a little luck. But nothing artificial!

Your nearby stream is teeming with life. If you are fortunate, you needn't go any farther to find an invaluable supply of materials for your new kind of fish collection. The object of an ecosystem is to create a state of biological balance without the help of mechanical devices. No artificial pumps, food, or filters are necessary; just the right amount of sunlight, fish, plants, animals, temperature, and water are needed. Since your stream or pond itself is most probably in biological balance in its natural state, the best idea is to copy that model in your fish tank, but on a smaller scale. Let nature take care of the rest.

It is quite obvious that fish, as living organisms, require food and oxygen. So the first task is to ensure an adequate supply of both. The key is the successful growth of plant-life in the tank. The plants will supply the oxygen to the fish through photosynthesis and will be a source of food. Therefore, place your tank where it can receive adequate sunlight. Once the plants begin to grow, you can add the animal life. The fish (any arbitrary collection of minnows will do) give off carbon dioxide which the plants use in photosynthesis.

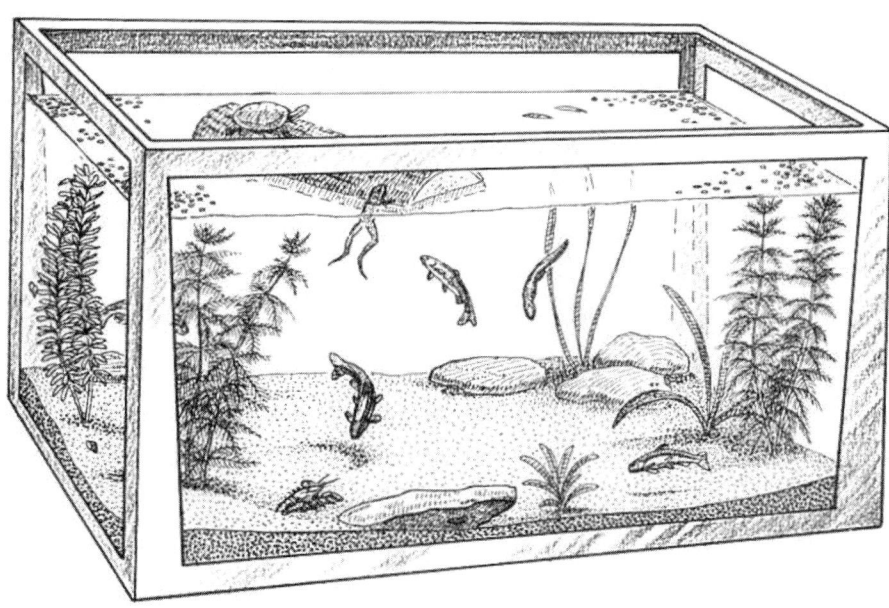

A balanced fresh-water aquarium is a miniature ecosystem.

If there are too many fish for the amount of plant life, the fish will die from the resulting imbalance. A few plants will give enough oxygen for only a few fish. Don't overcrowd the tank. A ten-gallon tank should not exceed ten fish, even in the most plentiful plant environment.

Now that the fish and plants are in oxygen and food balance, we need to deal with the waste and the cleaning of the tank. If you were fortunate in your pond sample, you may have caught some snails. These interesting little creatures will help clean the tank as they ingest tiny particles from the walls and floor of the tank. Remember, however, that they too must be considered as factors in the biological balance of the ecosystem. They too need the proper amounts of oxygen and food, so don't overcrowd the tank with snails.

Two more important factors in maintaining the success of the ecosystem are pH (acidity level) and temperature. As far as the pH factor is concerned, the only simple remedy is to periodically change the water and hope that your sample has a neutral pH value. Testing the water with litmus paper will indicate the acid and alkaline levels. The temperature level should be regulated to promote the growth of the plants. Of principal importance is maintaining the balance of the elements in the tank. If one element begins to dominate, the effects on the remaining elements will be immediately devastating.

If everything goes well, you will have a small aspect of nature's magic in your own home. Best assurance of success is an understanding of the interrelationships between organisms in the tank.

Materials:

- ten-gallon tank
- clean water (from stream, pond, or well)
- clean soil
- small sample of animal-life and plant-life (including snails)

Method:

Wash out tank to prevent undesirable bacteria. Bacteria are also important factors in maintaining balance. Certain strains of bacteria will aid in decomposing wastes and in supplying nitrogen necessary for plant growth, while others will directly attack and kill the fish. The safe way is to begin with a clean tank, clean water, and clean soil. It may be safe to assume that the water in the stream is clean. Of course if the stream has been tampered with by man, the balance may be upset. Part of the lesson of this experiment is to demonstrate the importance of the balance in a natural environment.

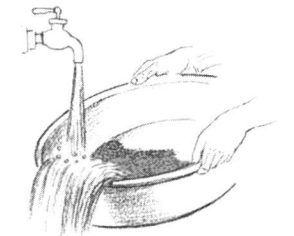

Bottom material should be clean.

Tank should be well cleaned.

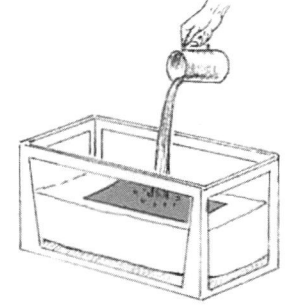

Floating paper or cardboard helps prevent stirring of bottom material while tank is being filled.

Collect the plants. Surface plants are usually abundant and easy to maintain. Gather some floating "clover-like" plants. A few anchored plants will supply plenty of oxygen, while some plankton and algae will be a good source of food. Let the plants grow for a while in the tank before adding the fish.

Next, you must net some fish. If you haven't a fish net handy, take an old stocking and stretch it over a bent clothes hanger. Use that as a net to collect your animal samples. Undoubtedly, you will have many species of living organisms in the sample. Most of the fun is trying to figure out what's in your tank! If you look closely at a small, moss-covered twig, it may move; in fact, it could very well not be a twig at all. It could be a home for some aquatic insect or a small crayfish-like organism. There are probably hundreds of tiny "things" swimming around; just look closely enough. Again, remember—they too are factors or elements in your ecosystem.

Now comes the challenge—trying to maintain the ecosystem successfully. "Successfully maintain" here means to promote the growth and enlargement of your system. This cannot be done if the balance of the system is disturbed. Most of the challenge lies in trying to keep alive all the individuals without using artificial means. If this can be done, the next step is to experiment by varying the numbers and kinds of individuals while still preserving the ecological balance. One example is the reproduction of some of the animals. Biologically speaking, new "mouths to feed" means new factors in our ecosystem. You may want to try enlarging the aquarium by adding some larger animals: frogs, water snakes, turtles, for example.

Two of the most important results of the experiment are the pride in your accomplishments and a renewed respect for the workings of nature. The latter may initiate or strengthen your views on conservation. The difficulty of restoring and preserving a balance in your small environment will demonstrate the severity of the problem of maintaining an ecological balance in the total environment. Your miniature ecosystem will give you an insight into ecology as a whole.

Article by Michael Andolina. Reprinted with permission from the June/July 1971 issue of New York State Conservationist magazine.

Streams and ponds furnish an abundance of material to stock the aquarium.

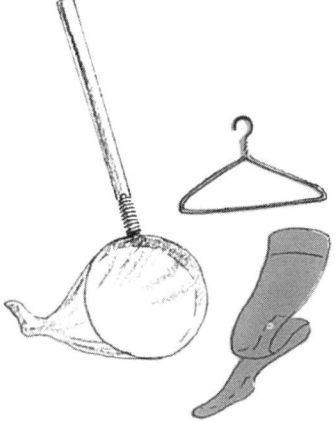

A simple collecting net can be made from a clothes hanger and a stocking.

The aquarium should be covered with glass.

A Miniature Ecosystem

Answer Key

Chapter 1

Workbook 1.3

A.
1. reproduction (or growth)
2. digestion
3. respiration
4. ingestion
5. absorption
6. irritability
7. transport
8. digestion
9. photosynthesis
10. excretion

B. *One possible answer:* Although salt crystals show growth, they do not display irritability. They do not ingest food, nor do they respire. The crystals do not excrete waste products. Therefore, salt crystals are not living.

Workbook 1.4

1. contents
2. glossary
3. appendix
4. body
5. glossary
6. title page
7. title page
8. contents
9. glossary
10. appendix
11. body

Workbook 1.5

A.
1. *Life Science* includes a title page, contents, body, appendix, and glossary.
2. The glossary begins on page 255.
3. The words in the glossary are arranged alphabetically.

B.
1. Title page: The first page of a book which includes the title, author, publisher, place and year of publication.
2. Contents: An outline of the book with page numbers for each chapter, found in the front of a book.
3. Body: The main part of the book.
4. Appendix: A part of some books containing charts, maps, or other detailed information, found in the back of the book.
5. Glossary: A micro-dictionary found in the back of some books.

Workbook 1.6

Formal Lab #1 Questions:
1. Answers will vary.
2. How much weight each plant could lift was the dependent variable.
3. One possible answer: Seedlings that were too weak to lift their cups tried to grow out from under the cups.

Extra credit: One possible answer: The large seeds were stronger than the small seeds. Dicots such as the sunflower seedlings tended to be stronger than monocots such as the corn seedlings, even though they were about the same size.

Chapter 2

Workbook 2.2

Compare the diagram to Figure 2.2 in the text. Check to make sure the diagram includes labels and a title.

Workbook 2.3

A. Mega, ---, ---, kilo, hecto, deka, basic unit, deci, centi, milli, ---, ---, micro

B.
1. meter
2. millimeter
3. milliliter
4. dekaliter
5. square centimeter
6. liter
7. centimeter
8. centigram
9. megagram
10. cubic centimeter
11. gram
12. decimeter
13. kilometer
14. micrometer
15. cubic centimeter

C.
1. 1
2. 100
3. 3.48
4. 5000
5. 26.1
6. 0.0001
7. 0.00001
8. 0.000056
9. 4000
10. 25
11. 20
12. 20
13. 20
14. 300
15. 4.5

Workbook 2.4

A.
1. 11.0 cm²
2. 5.5 cm²
3. 12.5 cm²
4. 27.5 cc

B.
1. 3.0 cm²
2. 3.0 cm²
3. 3.0 cc
4. 3.0 ml

C. 36,000,000 cc

Workbook 2.5

1. 1 cc is equal to 1 ml.
2. The formula or rectangular area is A = L × W.
3. The formula for rectangular volume is V = L × W × H.
4. The metric system is based upon the number ten.
5. Units must be included with the magnitude of a measurement.
6. "*Kilo-*" is the prefix that means 1000 times a unit.
7. "*Centi-*" is the prefix that means 1/100 of a unit.
8. One dekameter (1 dam) is larger.
9. The prefix "*micro-*" is the smallest in value.
10. The prefix "*centi-*" means "100."
11. *Possible answers include:* cent, centigrade, centipede, centurion, century.

Chapter 3

Workbook 3.1

1. *Student should list three of the following possibilities:* to learn; to enjoy; to preserve for later study; for the beauty of the specimens.
2. You should not collect poison ivy, poison oak, or poison sumac.
3. Only only two or three specimens should be taken from any one branch.
4. Ornamental plants require permission before collecting.
5. *Answers may vary. Possible answers include:* spiders, millipedes, centipedes, and "pill bugs."
6. Bees and wasps require caution when collected.
7. Four pieces of data which should be included with every specimen are the name of the specimen, when it was collected, who collected it, and where it was collected.
8. Glass containers should not be used to collect insects.

Chapter 4

Workbook 4.2

1. Hydrogen: one electron (e-) in first circle

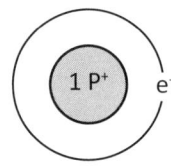

2. Chlorine: two e- in first circle, eight e- in second circle, seven e- in third circle

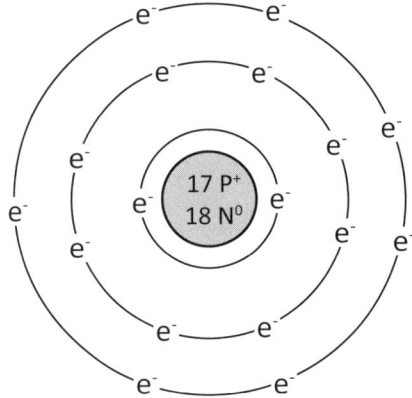

3. Nitrogen: two e- in first circle, five e- in second circle

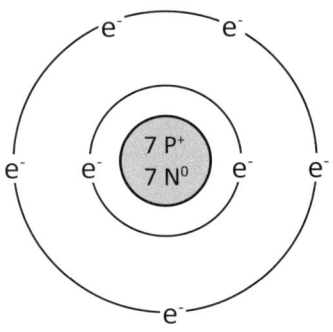

4. Oxygen: two e- in first circle, six e- in second circle

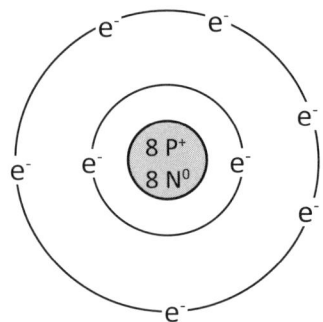

Workbook 4.3

1. no (or neutral)
2. +1 (or positive)
3. -1 (or negative)
4. oxygen
5. protein
6. hydrogen
7. black
8. oxygen
9. molecules
10. metals
11. 18

Workbook 4.4

C = 18.5%

H = 9.5%

O = 65.0%

N = 3.0%

Others = 4.0%

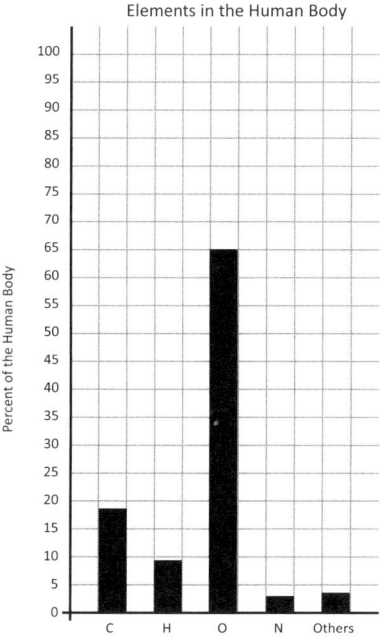

Workbook 4.5

1. electron, neutron, nucleus, atom, molecule, virus, human
2. The number of protons in an atom determines what kind of atom it is.
3. Most of an atom's volume is empty space.
4. The formula for water is H_2O.
5. There are three atoms in one molecule of water.
6. There are two elements in pure water.
7. *Any three of the elements on page 35 of the text. Do not accept formulas which include more than one element, such as NaCl, etc.*
8. Answers will vary. Possible answers (any three): H_2O, NaCl, HCl, CO_2, etc.
9. Answers will vary. Possible answers (any three): protoplasm, plasma, air, seawater.
10. Seawater is the one mixture on earth that is not in living things but is most like them.

Workbook 4.6

1. calcium
2. carbon
3. chlorine
4. copper
5. hydrogen
6. iodine
7. iron
8. magnesium
9. nitrogen
10. oxygen
11. phosphorus
12. potassium
13. sodium
14. sulfur
15. zinc

Chapter 5

Workbook 5.2

1. P
2. N
3. C
4. C
5. C
6. N
7. C
8. C
9. P
10. P
11. C
12. P
13. N
14. C
15. P
16. C
17. C
18. P
19. N
20. C
21. C

Workbook 5.3

1. Chloroplast and cell walls are found in plant cells but not in animal cells.
2. Centrioles are found in animal cells and not in plant cells.
3. Vacuoles are usually large in plant cells but small in animal cells.
4. The chromosomes hold the instructions for operating the cell.
5. The plasma membrane is also called the cell membrane.
6. The chloroplast is colored green.
7. Chloroplasts are needed in photosynthesis.
8. Ribosomes are the site where proteins are synthesized, or made.
9. The mitochondrion produces and releases energy to power the cell.
10. The lysosome is the area where food molecules are broken down.
11. Golgi bodies are for storing proteins.
12. The nuclear membrane controls entry to the nucleus.
13. Cyclosis is a circular movement.
14. Osmosis is the movement of water through a membrane.

Workbook 5.4

1. Robert Hooke looked at dead cork through a microscope and used the term "cells" to describe what he saw.
2. The word "cell" literally means "small room."
3. Diffusion is the movement of molecules from higher to lower concentration.
4. Energy is not expended in diffusion.
5. Diffusion is passive transport.
6. Osmosis is diffusion of water through a membrane.
7. *Any two of the following statements:* The cell is the unit of structure and function of organisms. Living things are made of cells. Cells are made by other living cells.
8. When your life began, your body consisted of one cell.

Workbook 5.5

1. nuclear membrane
3. instructions for the cell
4. cytoplasm

5. cell membrane, hold the cell together
6. mitochondria, release energy
7. endoplasmic reticulum
8. makes proteins
9. golgi body, storage
10. digestion
11. centrioles
12. chloroplast, photosynthesis
13. chlorophyll, capture sunlight

Workbook 5.6

Answers may vary. The concentration of water was lower inside the egg than outside of it, because egg white is a mixture of many other substances besides water. In osmosis, water moves through a semipermeable membrane from a higher concentration of water to a lower concentration. This is why the water in the glass moved through the egg's membrane and into the egg.

Workbook 5.8

Formal Lab #2 Questions:

1. *Answers may vary, but should be similar to the following:* Water entered the cells of the potato that was soaked in pure water, because the cytoplasm of cells contains many other substances besides water. This means that the concentration of water was lower inside the cells than outside of them, so water moved into the cells through osmosis.

2. *Answers may vary, but should be similar to the following:* Because of osmosis, water left the cells of the potato that was soaked in salt water. This is because the concentration of water is lower in very salty water than it is in the cytoplasm of cells, and in osmosis, water moves from a higher concentration to a lower concentration.

3. *Answers may vary, but should be similar to the following:* If we did not have a control with which to compare the other potatoes, we could not be sure if potatoes normally become more rigid or more limp when they are cut in half. Without a control, we would not have proof that the difference in firmness of the potatoes (the dependent variable) was due to the type of water they were soaked in (the independent variable).

Chapter 6

Workbook 6.2

The three basic necessities of life are energy, a way to use energy, and a way to excrete waste.

Workbook 6.3

The eight life functions are transport, irritability, nutrition, respiration, excretion, synthesis, growth, and reproduction.

Workbook 6.4

1. The only life function performed by viruses is reproduction.
2. A car cannot perform the life functions of growth, synthesis, reproduction, or irritability.
3. A flagellum is used for the life function of irritability.
4. Reproduction is the most outstanding characteristic of bacteria.
5. Binary fission is a type of reproduction.
6. Organisms in Kingdom Archaea thrive in extreme conditions.
7. <u>Volvox sp.</u> is a eukaryote.

Workbook 6.5

1. temperature
2. oxygen
3. moisture
4. temperature
5. light
6. food
7. moisture
8. oxygen and/or moisture
9. temperature
10. food

Workbook 6.6

1. DNA
2. virus
3. ATP
4. anaerobic
5. prokaryotes
6. blue-green algae
7. bacterial spore
8. pasteurization
9. mutualism
10. binary fission
11. flagellum
12. humus
13. parasitism
14. colony
15. heterotrophs

Workbook 6.7

One possible answer: Bacteria are needed as decomposers and as nitrogen fixers. Bacteria are useful in making foods such as cheese, pickles, vinegar, etc. Bacteria may be used to produce insulin and other medicines.

Workbook 6.8

Formal Lab #3 Questions:

1. *Answers will vary.*
2. *Answers will vary.*
3. *Answers will vary. Possible answers include:* 1) There may have been bacteria on the q-tip, on the petri dishes, or on the mouth of the bottle of liquid nutrient agar. These sources of error could be eliminated by sterilizing these items. 2) One of the dishes may have been left uncovered longer than the others, and thus collected more bacteria from the air. This source of error could be eliminated by more attention to standardizing the procedure. 3) The bacteria could have been collected differently from each surface. This source of error could be eliminated by more attention to standardizing the procedure. 4) The colonies of bacteria in one of the dish may have been counted inaccurately. This could be corrected by greater care or by more experience in distinguishing one colony from another.
4. Bacterial colonies on parts of the agar which were not touched by the q-tip probably came from bacteria in the air (or possibly from bacteria in the liquid agar or on the petri dish).
5. Bacterial colonies in the petri dish labeled "Control" probably came from bacteria in the air (or possibly from bacteria in the liquid agar or on the petri dish).

Chapter 7

Workbook 7.2

1. The science of naming creatures is called taxonomy.
2. The taxon species contains just one kind of organism.
3. Binomial nomenclature is the term for using two names to identify an organism.
4. The two taxons used in a scientific name are the genus and the species.
5. The largest of the seven taxons is the kingdom.
6. There are only six kingdoms of living organisms.
7. Families are subdivisions of orders.
8. Creatures which are almost, but not exactly the same, would be in all the same groups except for their species.
9. A kingdom could contain organisms that are most different.
10. Different protists in the same phylum must also belong to the same kingdom.

Workbook 7.3

1. Aristotle started the first orderly classification of living things.
2. The most important thing Carolus Linnaeus did for taxonomy was inventing the seven-taxon system.
3. The house fly and the human are in the same kingdom.
4. A red oak and an unborn baby cannot be in the same species.
5. To write scientific names correctly, you should underline both the genus and the species, capitalize the genus, and keep the species in lowercase.
6. My scientific name is <u>Homo sapiens</u>.
7. Scientists use structure as the basis for classifying organisms.
8. Two organisms in the same genus are very similar.

Workbook 7.4

Compare the diagram to Figure 7.14 in the text. Check to make sure the diagram includes labels and a title.

Workbook 7.5

1. Kingdom Protista has unicellular organisms with organelles.
2. Slime mold and golden algae belong to the same kingdom.
3. Euglena, amoeba, and paramecium belong to the same kingdom.
4. The amoeba moves by means of a pseudopod.
5. The paramecium has cilia for locomotion.
6. The euglena has a flagellum to help it perform the life function of irritability.
7. The euglena is both plant-like and animal-like.
8. The amoeba can reproduce by binary fission.
9. All protists have a nucleus, with a membrane, that controls the cell.
10. The oak tree and the paramecium share none of the same taxons.
11. All the phyla of the Kingdom Protista have organelles, a nuclear membrane, and unicellular bodies.

Chapter 8

Workbook 8.2

Fungi are eukaryotes with nuclear membranes and organelles, unlike archaeans and bacteria, which are prokaryotes without nuclear membranes or organelles.

Workbook 8.3

Fungi have specialized cells while protists do not. Most fungi have chitinous cell walls, but protists do not.

Workbook 8.4

A. *Four of the following characteristics:* Fungi 1) possess nuclear membranes and organelles; 2) have chitinous cell walls; 3) are heterotrophic; 4) are multicellular with specialized cells; 5) take in nutrients digested externally.

B. *Answer should describe five different ways fungi are helpful or harmful. One possible answer:* Fungi are helpful 1) in making bread and wine; 2) in producing penicillin and other medicines; 3) as decomposers that recycle nutrients; 4) as food. Fungi can also be harmful; 5) some are poisonous; 6) some cause athlete's foot or other diseases.

Workbook 8.5

1. fungi
2. mycologist
3. heterotrophic
4. annulus
5. stipe
6. hyphae
7. chitin
8. ascus
9. basidia

Workbook 8.6

Compare the diagram to Figure 8.6 in the text. Check to make sure the diagram includes labels and a title.

Chapter 9

Workbook 9.2

1. Plants have nuclear membranes and organelles, but prokaryotes do not.
2. Plants are multicellular with specialized cells; protists are not.
3. Plants have cellulose cell walls and chlorophyll; fungi have chitinous cell walls.
4. Plants have cell walls and chloroplasts; animals do not.

Workbook 9.3

Compare the diagram to Figure 9.20 in the text. Check to make sure the diagram includes labels and a title.

Workbook 9.4

1. Autotroph means "self-feeder."
2. In photosynthesis, plants use light energy to make food.
3. Plants have nuclear membranes; archaeans, bacteria, and/or prokaryotes do not.
4. Plants have cellulose cell walls; fungi do not.
5. Plants have chlorophyll but fungi do not.
6. Phylum Chlorophyta is the phylum of green algae.
7. Spirogyra sp. has a spiral chloroplast.
8. Starch, in the pyrenoids, turns iodine blue-black.
9. Lichens, containing both algae and fungi, are a good example of mutualism.
10. Lichens are organisms that can live on bare rocks.

Workbook 9.5

Ferns grow where it is moist because they need moisture to reproduce. Ferns are taller than mosses because they have vascular systems.

Workbook 9.6

1. Bryophyta
2. vascular system
3. reproduction
4. rhizoids
5. Plantae
6. Bryophyta
7. Musci
8. Polytrichum
9. commune
10. <u>Polytrichum commune</u>

Workbook 9.7

1. Plantae
2. Tracheophyta
3. vascular system
4. frond
5. rhizome
6. sori
7. sporangia
8. Filicineae
9. spores
10. autotrophs

Workbook 9.8

1. All tracheophytes have a vascular system.
2. Class Filicineae contains plants that do not produce seeds.
3. The root word "*gymno*" means "naked."
4. Plants in Class Angiospermae have flowers.

5. Class Gymnospermae contains the conifers.
6. Class Angiospermae is also called the class of flowering plants.
7. In the flowering plants, the seeds are hidden inside of fruit.
8. Two important uses of conifers are for lumber and for paper.
9. "Monocotyledonae" means "one cotyledon" (seed leaf).
10. Dicots are angiosperms.

Workbook 9.10

Compare the diagram to Figure 5.6 in the text. Check to make sure the diagram includes labels and a title.

Chapter 10

Workbook 10.2

1. atoms
2. molecules
3. organelles
4. cells
5. tissues
6. organs
7. systems
8. organisms

Workbook 10.3

Guard cells in the lower epidermis on the undersides of leaves control how rapidly transpiration occurs. Their sausage-like shape with a thickened side toward the stomate allows them to bend when turgid. When flaccid, they close the stomate. The stomate opens into the spongy layer where water vapor moves out of the loosely packed cells and into the air.

Workbook 10.4

A.

1. tissue
2. fibrous root
3. tap root
4. phloem
5. root hairs
6. herbaceous stems
7. bark
8. bulb
9. palisade layer
10. stomates
11. transpiration
12. xylem
13. auxins
14. phototropism

B. *In any order:*

1. phototropism, light
2. geotropism, earth
3. hydrotropism, water
4. thigmotropism, touch
5. chemotropism, chemicals

Workbook 10.5

Compare the diagram to Figure 10.6 in the text. Check to make sure the diagram includes labels and a title.

Workbook 10.6

1. abiotic
2. prokaryotes
3. Protista
4. Animalia
5. Fungi
6. Chlorophyta
7. Bryophyta
8. Tracheophyta
9. ferns
10. Gymnospermae
11. Angiospermae
12. Monocotyledonae
13. corn, lily, onion, or any other monocot
14. Dicotyledonae
15. peanut, apple, celery, or any other dicot

Workbook 10.7

Compare the diagram to Figure 10.14 in the text. Check to make sure the diagram includes labels and a title.

Chapter 11

Workbook 11.2

A. *One possible answer:* Asexual reproduction involves only one parent and usually produces offspring more quickly than sexual reproduction. Sexual reproduction involves two parent cells and their fusion allows for variation within a species.

B. egg cell forms, sperm cell forms, pollination, fertilization, zygote grows, dispersal, germination

Workbook 11.3

1. sporulation
2. asexual reproduction
3. vegetative propagation
4. corolla
5. stamen
6. pistil
7. sexual reproduction
8. gametes
9. zygote
10. cross-pollination
11. seed dispersal
12. germination

Workbook 11.4

1. F
2. B
3. S
4. V
5. B
6. V
7. V
8. F
9. V
10. S

Workbook 11.5

Compare the diagram to Figure 11.9 in the text. Check to make sure the diagram includes labels and a title.

Workbook 11.6

Compare the diagram to Figure 11.10 in the text. Check to make sure the diagram includes labels and a title.

Chapter 12

Workbook 12.2

1. Bluegills and humans are not in the same order.
2. The pumpkinseed is most like the bluegill.
3. Since the gray seal is in the same class as the bottlenose dolphin, it must also be in the same phylum.
4. Only one kind of organism can be in the Genus Tursiops and species truncatus.
5. The scientific name of a blue whale is <u>Balaenoptera musculus</u>.

Workbook 12.3

Compare the diagram to Figure 12.25 in the text. Check to make sure the diagram includes labels and a title.

Possible explanation: Complete metamorphosis is the change of an insect through four distinct stages: egg, larva, pupa, adult.

Workbook 12.4

Compare the diagram to Figure 12.23 in the text. Check to make sure the diagram includes labels and a title.

Workbook 12.5

A. *In any order:* Animals are multicellular eukaryotes; animals are heterotrophic (that is, they ingest food); animals have no cell walls.

B.

1. sessile
2. pores
3. pupa
4. radial symmetry
5. regeneration

Workbook 12.6

1. Porifera
2. azure vase sponge *(Answers may vary.)*
3. The skeletons of sponges can be used for cleaning.
4. Cnidaria
5. hydra *(Answers may vary.)*
6. radial symmetry
7. two tissue layers
8. flatworms
9. Planaria *(Answers may vary.)*
10. bilateral symmetry
11. three tissue layers
12. Nematoda
13. pork worm *(Answers may vary.)*
14. complete digestive system
15. segmented worms
16. hermaphroditic
17. breathe through skin
18. Mollusca
19. oysters, squids, snails *(Answers may vary.)*
20. soft-bodied with a muscular foot
21. Arthropoda
22. insects, crustaceans *(Answers may vary.)*
23. chitinous exoskeleton
24. *Possible answers include:* pests, pollinators, food, disease vectors.
25. Echinodermata
26. starfish *(Answers may vary.)*
27. spiny skin
28. radial symmetry
29. regeneration

Chapter 13

Workbook 13.2

1. omnivore
2. insectivore
3. Chordata
4. gills
5. cold-blooded
6. placenta
7. estivate

Workbook 13.3

Answers may vary. The placenta is a part of an unborn mammal. It is soft and spongy because it is full of blood vessels. Cows, cats, and dogs, as well as humans, use a placenta to live inside their mothers. Marsupials like the opossum, kangaroo, and koala do not have placentas. The spiny anteater and duck-billed platypus do not need a placenta because they lay eggs.

Workbook 13.4

Compare the diagram to Figure 5.8 in the text. Check to make sure the diagram includes labels and a title.

Workbook 13.5

1. Animalia is the name of a kingdom.
2. Annelida, Arthropoda, Chordata, Cnidaria, Echinodermata, Mollusca, Nematoda,

Platyhelminthes, and Porifera are names of phyla.
3. Vertebrata is the name of a subphylum.
4. Agnatha, Amphibia, Aves, Chondrichthyes, Mammalia, Osteichthyes, and Reptilia are classes.
5. Animalia is a group which includes all the others.

Workbook 13.6

1. Phylum Porifera
2. Phylum Porifera
3. all of these phyla
4. none of these phyla
5. Phylum Cnidaria
6. Phylum Platyhelminthes
7. none of these phyla
8. all of these phyla
9. Phylum Nematoda
10. none of these phyla
11. Phylum Annelida
12. Phylum Annelida
13. none of these phyla
14. Phylum Mollusca
15. all of these phyla
16. Phylum Arthropoda
17. Phylum Arthropoda
18. Phylum Echinodermata
19. Phylum Chordata
20. Phylum Arthropoda
21. Phylum Chordata
22. Phylum Chordata
23. Phylum Chordata
24. Phylum Porifera
25. Phylum Chordata
26. Phylum Arthropoda
27. all of these phyla
28. Phylum Arthropoda
29. none of these phyla
30. none of these phyla
31. Phylum Chordata
32. Phylum Cnidaria
33. Phylum Platyhelminthes
34. Phylum Nematoda
35. Phylum Annelida
36. Phylum Arthropoda
37. Phylum Chordata
38. Phylum Chordata
39. Phylum Chordata

Midway Review

Part II. Measurements and Chemistry

1.
   - Define the question or problem.
   - Collect data through research.
   - Formulate a hypothesis to answer the question.
   - Experiment using a control and a variable group.
   - Interpret the results of the experiment and draw a conclusion that supports or rejects the hypothesis.
   - Share your data.
2. 500 cm, 0.847 l, 8 ml
3. C=carbon, H=hydrogen, O=oxygen, N=nitrogen, S=sulfur, P=phosphorus, I=iodine, Na=sodium, K=potassium, Fe=iron, Cu=copper, Cl=chlorine, Mg=magnesium, Ca=calcium, Zn=zinc
4. Water: H_2O, Oxygen gas: O_2, Table salt: NaCl, Carbon dioxide: CO_2

Part III. Cells

1. transport, irritability, nutrition, respiration, excretion, synthesis, growth, reproduction

   Cell wall: provides structure
   Centrioles: assist in cell division in animals
   Chloroplast: organelle where photosynthesis occurs to capture energy from sunlight
   Chromosomes: contain the information to operate the cell
   Endoplasmic reticulum: provides a surface for chemical reactions
   Golgi body: transports, modifies, and stores proteins
   Lysosome: contains digestive enzymes
   Mitochondrion: organelle where cellular respiration occurs to release energy
   Nucleus: controls cell activity
   Ribosomes: make proteins
   Vacuoles: contain air or food

2. Osmosis is a special type of diffusion in which water passes through a cell membrane. The water molecules that pass through the cell membrane in osmosis move from a higher concentration of water to a lower concentration of water.

Part IV. Classification

1. kingdom, phylum, class, order, family, genus, species
2. Archaea, Bacteria, Protista, Fungi, Plantae, Animalia
3. The three basic shapes of bacteria are cocci (round), bacilli (rod), and spirilla (spiral).
4. Phylum Chlorophyta, ex: green algae, brown algae, etc.; Phylum Bryophyta, ex: mosses, hornwort, etc.; Phylum Tracheophyta, ex: ferns, daisies, cedars, etc.
5. Class Filicineae, ex: common bracken and other ferns; Class

Gymnospermae, ex: pines, cedars, and other conifers; Class Angiospermae, ex: grass, oaks, corn, peas, tulips, and other flowering plants

6. The two subclasses of angiosperms are Subclass Monocotyledonae and Subclass Dicotyledonae. Monocots have one seed leaf (cotyledon), parallel veins, flower parts in multiples of 3 or 4, and scattered vascular bundles. Dicots have two seed leaves (cotyledons), branching veins, flower parts in multiples of 4 or 5, and vascular bundles in a ring.

7. **Porifera** (ex: sponges)
   Cnidaria (ex: jellyfish, hydra)
   Platyhelminthes (ex: flatworms, beef tapeworm, Planaria)
   Nematoda (ex: roundworms, hookworm, pork worm)
   Annelida (ex: earthworms)
   Mollusca (ex: snails, octopuses)
   Arthropoda (ex: insects, lobsters, spiders)
   Echinodermata (ex: starfish, sand dollars, sea urchins)
   Hemichordata (ex: acorn worms)
   Chordata (ex: fish, dogs, humans)

8. **Class Agnatha** contains jawless animals with gills. **Class Chondrichthyes** contains animals with a cartilaginous skeleton. **Class Osteichthyes** contains animals with gills, scales, and a bony skeleton. **Class Amphibia** contains animals with gills and then lungs. **Class Reptilia** contains animals with lungs and a scaly skin. **Class Aves** contains warm-blooded animals with feathers. **Class Mammalia** contains warm-blooded animals with hair and mammary glands.

Part V. Reproduction, Growth, and Response

1. *Student should identify four of the following:* regeneration, binary fission, budding, conjugation, sporulation, vegetative propagation (cuttings, grafting, runners, layering, bulbs and tubers, etc.).

2. Mitosis is the process of normal cell division in which two new cells are formed, each with a complete set of chromosomes. The five stages of mitosis are interphase, prophase, metaphase, anaphase, and telophase.

3. *One possible answer*: Asexual reproduction involves only one parent and usually produces offspring more quickly than sexual reproduction. Sexual reproduction involves two parent cells and their fusion allows for variation within a species.

4. Sunlight causes the auxin in cells to move to the shaded side of the plant. The shaded side of a plant stem has a greater concentration of auxin, so they grow longer than the cells on the sunny side. This unequal growth causes the plant to grow toward the light.

Chapter 14

Workbook 14.2

1. fat
2. amino acids
3. glucose
4. monosaccharide
5. nutrient
6. proteins
7. cellular respiration
8. chemosynthesis

Workbook 14.3

1. Hydrogen oxide is another name for water.
2. A few types of amino acids make up thousands of kinds of proteins.
3. Iron is the mineral in hemoglobin that carries oxygen.
4. Carbohydrates are made of carbon, oxygen, and hydrogen, with two hydrogens for every oxygen.
5. Sucrose is a disaccharide.
6. Glycogen is a starch made from glucose molecules.
7. Sodium and chlorine are minerals in table salt.
8. Ascorbic acid is another name for vitamin C.
9. Carotene is another name for the vitamin A in carrots.
10. Milk and milk products are a good source of calcium, a mineral needed for strong bones.
11. ATP is a molecule used to hold energy.
12. Glucose is combined with oxygen in cellular respiration.
13. Chlorophyll is needed for photosynthesis.
14. In photosynthesis, oxygen (or water) is given off as a waste.
15. Human beings are heterotrophs.

Workbook 14.4

A.

Food examples and order of nutrient groups may vary.

water; water

fat; bacon

protein; hamburger

carbohydrate; spaghetti

vitamins and minerals; broccoli

B.

$C_6H_{12}O_6 + 6O_2 \longrightarrow 6H_2O + 6CO_2 +$ Energy

C.

$6CO_2 + 12H_2O +$ Light Energy $\longrightarrow$ $C_6H_{12}O_6 + 6O_2 + 6H_2O$

Workbook 14.5

One possible answer: The autotrophs carry out photosynthesis, most by using chlorophyll to capture sunlight. But both plants and heterotrophs must carry out respiration in which the monosaccharide glucose is combined with oxygen to release energy. This energy is then stored in adenosine triphosphate (ATP) molecules.

Workbook 14.6

1. Vitamin C
2. Calcium
3. Sodium
4. Vitamin C
5. Iron
6. Iodine
7. Vitamin B_1
8. Sodium
9. None of these
10. None of these

Workbook 14.7

1. C
2. W
3. P
4. C
5. P
6. P
7. V
8. C
9. P
10. P

Workbook 14.8

Descriptions may vary.

A. fats and oils
B. proteins
C. carbohydrates
D. vitamins and minerals

Workbook 14.9

Formal Lab #4 Questions:

1. The amount of carbon dioxide produced by the yeast depends on the amount of sugar the yeast is assimilating.
2. The independent variable is the amount of sugar in each bottle.

Chapter 15

Workbook 15.2

1. 5 cal
2. 1 °C
3. 300 cal
4. 3 °C
5. 6 cal
6. 17 °C
7. 8 °C
8. 2 °C
9. 2 g
10. 25 °C

Workbook 15.3

1. 92%
2. 85%
3. 24 kg

Workbook 15.4

1. Rub sample on paper towel or napkin. A translucent spot that does not evaporate indicates the presence of fat.
2. *Either of the following tests for protein:*
   a) Place equal amounts of water and sample in a test tube and shake. Add sodium hydroxide (NaOH) solution and stir. Add copper sulfate ($CuSO_4$) solution. A color change to violet indicates the presence of proteins.
   b) Add sample to a small amount of nitric acid, boil, pour off the acid and rinse with water. Cover sample with ammonium hydroxide (NH_4OH). A color change to orange indicates the presence of protein.
3. Add a few drops of silver nitrate ($AgNO_3$) solution to the sample. A white precipitate forms when chlorine is present.
4. Blue cobalt chloride ($CoCl_2$) paper turns white in the presence of water vapor.

Chapter 16

Workbook 16.2

1. The term "organism" includes all the other terms.
2. The cell is the unit of structure and function in living things.
3. A system is a group of organs working together to carry out a life function.
4. Epidermis, nerves, and muscles are examples of tissue.
5. Together, the mouth, esophagus, stomach, intestines, and associated organs are an example of a system.

Workbook 16.3

Any order:

transport: circulatory

irritability: muscular, skeletal, and/or nervous

nutrition: digestive

respiration: respiratory

excretion: excretory

synthesis: endocrine

reproduction: reproductive

Workbook 16.4

Compare the diagram to Figure 16.4 in the text to make sure the regions of the body are labeled correctly.

Workbook 16.5

1. D
2. E
3. F
4. B
5. A
6. G
7. C

Workbook 16.6

Refer to Chapter 16, pgs. 149–151, in text.

Chapter 17

Workbook 17.2

1. I
2. V
3. I
4. V
5. V
6. I
7. V
8. I
9. V

Workbook 17.3

1. joint
2. cartilage
3. striations
4. involuntary
5. adipose cells
6. cardiac muscle
7. ligaments
8. tendon
9. marrow
10. bone

Workbook 17.4

A.

1. cranium (skull)
2. vertebrae (neck bone)
3. rib (rib cage)
4. ilium (hip bone)
5. femur (thigh bone)

B.

Any order: Provide support; protect vital organs; provide storage of minerals; act as levers to which muscles are attached; make blood cells.

Workbook 17.5

Compare the diagram to Figure 17.3 in the text. Check to make sure the diagram includes labels and a title.

Workbook 17.6

Compare the diagrams to Figure 17.7 in the text. Check to make sure the diagrams have titles.

Chapter 18

Workbook 18.2

A. Mechanical digestion produces smaller pieces of the same material; chemical digestion results in pieces of simpler chemical composition.

B. *Any order:*

- pepsin, stomach, proteins
- amylase, pancreas and/or salivary glands, starches
- trypsin, pancreas, proteins
- lipase, pancreas, lipids (fats)
- maltase, small intestine, maltose
- sucrase, small intestine, sucrose

Workbook 18.3

Compare the diagram to Figure 18.9 in the text. Check to make sure the diagram includes labels and a title.

Workbook 18.4

1. villus
2. cuspid
3. amylase
4. pepsin
5. bile
6. peristalsis
7. lacteal
8. liver
9. gall bladder
10. esophagus
11. pyloric sphincter
12. tongue
13. large intestine
14. pancreas
15. rectum

Workbook 18.5

A.

1. mouth
2. amylase, maltase, sucrase
3. glucose
4. small intestine
5. lipase
6. fatty acids, glycerol
7. stomach
8. pepsin, trypsin
9. amino acids

B.

1. incisors, cutting
2. cuspids, ripping meat
3. bicuspids, grinding
4. molars, crushing food

Workbook 18.6

1. The thick muscle in the mouth is the tongue.
2. The bumps on the tongue are called taste buds.
3. The four basic tastes are sweet, salt, sour, and bitter.
4. An adult human has 32 teeth.
5. The small intestine is the longest and most important organ of the digestive system.
6. The body uses digested foods for energy, growth and repair, and metabolic regulation.

7. The wave-like movements of the digestive system are called peristalsis.
8. Capillaries and lacteals are inside villi.
9. Food remains in the stomach for approximately 2–5 hours.
10. The valve between the stomach and the small intestine is called the pyloric sphincter. It controls exit from the stomach.
11. Vitamins B and K are two chemicals made by bacteria in the large intestine.

Workbook 18.7
Formal Lab #5 Questions:

1. The experiment shows that starch molecules are broken down into sugar molecules in the mouth. The fact that the rice which was chopped with a knife tested negative for sugar shows that it was the saliva which broke down the starch molecules, not the crushing action of the teeth.
2. By testing saliva and rice individually for simple sugars, we learned that neither of them contains sugar on their own. Thus, when the chewed rice tested positive for sugar, we knew that a mechanical or chemical change had taken place in the rice and saliva mixture. If we had not tested the saliva individually for sugar, we would not know whether the chewed rice tested positive for sugar only because saliva naturally contains sugar.

Chapter 19

Workbook 19.2
Compare the diagram to Figure 19.3 in the text. Check to make sure the diagram includes labels and a title.

Workbook 19.3
1. Waste material from a cell diffuses through the lymph fluid into capillaries.
2. Oxygen (O_2) is carried by hemoglobin in the red blood cells.
3. In systemic circulation, oxygen-poor blood returns to the heart through veins.
4. The veins have valves.
5. Flowing through the inferior vena cava, the blood enters the right atrium.
6. The right ventricle pumps blood to the lungs.
7. Pulmonary veins contain oxygen-rich blood.
8. The left ventricle is the strongest chamber of the heart.
9. The large artery that curves out of the top of the heart to carry blood to the body is the aorta.
10. Each artery is thick-walled and muscular.
11. Plasma is the liquid part of the blood.
12. White blood cells fight disease organisms.
13. Fibrinogen is a blood protein needed for clotting.
14. Systemic circulation takes blood to the body.
15. Cardiac muscle is striated and has one nucleus in each cell.

Workbook 19.4
1. plasma
2. yellowish to clear liquid
3. platelets
4. help in clotting
5. fight disease
6. red blood cells
7. carry oxygen
8. veins
9. thin-walled with valves
10. arteries
11. carry blood away from the heart
12. heart
13. pump blood
14. capillaries

Workbook 19.5
A. *One possible answer:* Oxygen-rich red blood in the left atrium moves into the left ventricle and is pumped through the aorta into smaller arteries and finally through the tiny capillaries. In the capillaries, the hemoglobin in the red blood cells releases the oxygen. By diffusion, oxygen moves out of the capillaries and into the surrounding cells while carbon dioxide moves in the opposite direction. The oxygen-poor "blue" blood moves into the veins which join to form the superior vena cava and the inferior vena cava. The blood is returned to the right atrium of the heart.

B. *One possible answer:* Oxygen-poor blood is squeezed through the right atrium and pumped by the right ventricle through the pulmonary arteries to the lungs. Carbon dioxide diffuses from the capillaries into the air sacs, and oxygen moves from the air sacs into the capillaries. The oxygenated "red" blood moves through the pulmonary veins back to the left atrium of the heart.

Chapter 20

Workbook 20.2

Compare the diagram to Figure 20.6 in the text. Check to make sure the diagram includes labels and a title.

Workbook 20.3

1. The functions of the respiratory system are to exchange gases and regulate body temperature.
2. Oxygen is required by every living cell in the human body.
3. Carbon dioxide must be eliminated from every cell.
4. An ideal respiratory surface is moist, thin, and close to oxygen.
5. Fish use gills for respiration.
6. The respiratory surface in worms is their skin.
7. An amoeba uses its cell membrane for respiration.
8. Alveoli are clusters of air sacs. Each air sac is filled with air on the inside and surrounded by capillaries on the outside.
9. External respiration is the exchange of gases in the lungs.
10. Cellular respiration is the oxidation of glucose in a cell to release energy with carbon dioxide as a waste product.

Workbook 20.4

1. vocal cords
2. nostrils
3. diaphragm
4. alveolus
5. bronchus
6. mucus
7. expiration
8. hiccups
9. epiglottis
10. oxidation

Workbook 20.5

A.
1. 7
2. 2
3. 1
4. 8
5. 4
6. 10
7. 9
8. 5
9. 3
10. 6

B. *One possible answer:* The skeletal system protects the circulatory and respiratory organs, and the muscles move the rib cage during external respiration. The skeletal system also makes red blood cells. In return, the bone and muscle cells use the oxygen that diffuses into the blood in the lungs and is pumped through the body by the heart.

Workbook 20.6

1. The function of the human respiratory system is to exchange gases.
2. The nose and mouth moisten the air before it reaches the lungs.
3. The pharynx is the space at the back of the nose and mouth.
4. The epiglottis prevents food from entering the trachea.
5. Vibrations of the vocal cords cause sounds for speech.
6. The trachea directly connects the pharynx to the bronchi.
7. The trachea has rings of cartilage to keep it open.
8. The bronchioles end in alveoli.
9. Like root hairs and villi, the air sacs or alveoli greatly increase the surface area of an organ.
10. When too much carbon dioxide is in the blood, the brain tells the diaphragm and rib muscles to contract.
11. Pulmonary circulation brings blood to the lungs.
12. The process of oxygen moving from the blood into the cells for use in oxidation is called cellular respiration.
13. Waste carbon dioxide moves from the cells to the lymph fluid and into the blood by diffusion.
14. The veins and diaphragm are made of smooth muscle.
15. Expiration occurs when the rib muscles and the diaphragm relax.

Chapter 21

Workbook 21.2

1. excretory
2. lungs
3. mouth
4. nostrils
5. solid wastes
6. anus
7. liver
8. urea
9. bile pigments
10. kidneys
11. water
12. nitrogenous wastes
13. salts
14. skin
15. pores

Workbook 21.3

Compare the diagram to the cross-section of a kidney in Figure 21.3 in the text. Check to make sure the diagram includes labels and a title.

Workbook 21.4

1. nephrons
2. dialysis
3. nephron tubules
4. sweat glands
5. dermis
6. epidermis
7. urine
8. urea
9. ureters

Workbook 21.5

Carbon dioxide is produced as a waste product in the oxidation of glucose in every living cell. By diffusion the carbon dioxide moves out of the cells, through the lymph fluid, and into the capillaries. The blood carries the carbon dioxide to the right atrium, through the right ventricle and the pulmonary arteries, to the lungs. By diffusion the carbon dioxide moves into the air sacs and leaves the body through the mouth and/or nostrils during expiration. If the carbon dioxide is not eliminated from the body, the organism dies of asphyxiation.

Workbook 21.6

Compare the diagram to Figure 21.6 in the text. Check to make sure the diagram includes labels and a title.

Chapter 22

Workbook 22.2

A. *Two of the following glands:* salivary glands, sweat glands, mammary glands, liver (or accept gall bladder)

B. Unlike the other glands of the body, the endocrine glands do not have ducts.

Workbook 22.3

1. pituitary
2. thyroid
3. thyroid
4. thyroid
5. pituitary
6. parathyroid
7. pancreas
8. adrenal
9. thymus
10. pituitary
11. adrenal
12. gonads
13. pancreas
14. pancreas
15. adrenal
16. thyroid

Workbook 22.4

Five of the following:

Gland	Location
pituitary	center, base of the brain
thyroid	in the neck below the larynx
parathyroid	on the thyroid gland
thymus	in the chest
Islands of Langerhans	in the pancreas
adrenal	on top of each of the kidneys
testes	in the scrotum
ovaries	inside the lower abdomen

Workbook 22.5

Five of the following:

Disease	Cause	Hormone
acromegaly	overactive pituitary in adults	growth hormone
cretinism	underactive thyroid in a child	thyroxin
diabetes	underactive Islands of Langerhans	insulin
dwarfism	underactive pituitary in childhood	growth hormone
giantism	overactive pituitary in childhood	growth hormone
goiter	lack of iodine	thyroxin

Chapter 23

Workbook 23.2

Any order:

involuntary actions: medulla

balance: cerebellum

data input: five sense organs

simple reflex: spinal cord

messages: nerve cells

memory: cerebrum

voluntary actions: cerebrum

thinking: cerebrum

Workbook 23.3

Compare the diagram to Figure 23.3 in the text. Check to make sure the diagram includes labels and a title.

Workbook 23.4

1. Motor
2. Sensory
3. Associative
4. Sensory
5. Motor
6. Associative
7. Sensory
8. Motor
9. Associative
10. motor

Workbook 23.5

A. *Any order:*

cerebrum: thinking, sight, speech, voluntary actions

cerebellum: coordination and balance

medulla: involuntary actions (i.e., digestion, heartbeat, etc.)

B. In any order:

Organ	Sense	Nerves or Parts
eye	sight	optic nerve
ear	hearing	auditory nerve
nose	smell	olfactory nerve
tongue	taste	taste buds
skin	touch	pressure, pain, touch, heat, cold receptors

Answer Key

Workbook 23.6

1. sclera
2. lens
3. cornea
4. pupil
5. iris
6. retina
7. optic nerve
8. vitreous humor

Workbook 23.7

One possible answer: If the eye focuses light too close to its lens, **nearsightedness** is the result. The **rod** and **cone cells**, or **sensory neurons**, on the **retina** receive the light out of focus and the **optic nerve** sends a blurry image to the brain. When the eye focuses light too far from the lens, **farsightedness** is the result.

Workbook 23.8

1. optic nerve
2. retina
3. dendrites
4. cochlea
5. cone cells
6. convolutions
7. synapse
8. cerebellum
9. cerebrum
10. medulla
11. eustachian tube
12. farsightedness
13. iris
14. olfactory nerve
15. semicircular canals
16. auditory nerve
17. axon

Workbook 23.9

Formal Lab #6 Questions:

1. Answers may vary, but usually the experiment will indicate that both the student and his partner have the most neurons in their palms.
2. Different parts of the body need more neurons than others because they are designed for different jobs. For instance, an important job of the hands is gathering information about objects through the sense of touch. This job requires a lot of sensory neurons. Shoulders and calves do not need to be as sensitive because their main job is to move the body or other objects from place to place.

Chapter 24

Workbook 24.2

Answer should include the following information: In 1796, Edward Jenner used cow pox on James Phipps to vaccinate him against smallpox.

Workbook 24.3

1. The two basic types of disease are infectious and noninfectious diseases.
2. There are four types of noninfectious diseases: deficiency diseases, environmental diseases, functional disorders, and genetic diseases.
   Examples: **Deficiency:** any disease in Figure 24.3 in text.
   Environmental: lung cancer, skin cancer, diseases of the liver.
   Functional disorder: any disease in Figure 24.4 in text.
   Genetic: color blindness, hemophilia, sickle cell anemia, Down syndrome.
3. Infectious diseases can be spread by droplets, food, contact, cuts, and vectors.
4. The human body's three lines of defense against disease are the skin (epidermis); gastric juice (hydrochloric acid) and phagocytes; and antibodies.
5. The four functions of antibodies are neutralization, lysis, opsonization, and agglutination.

Workbook 24.4

1. The bacterium must be present in every case of the disease.
2. The bacterium must be grown in a pure culture.
3. Bacteria from the pure culture must cause the disease in a healthy organism.
4. The bacterium must be reisolated, cultured, and identified as identical to the original.

Workbook 24.5

1. puncture, tetanus
2. hemophilia, noninfectious
3. vector, malaria
4. host, beef tapeworm
5. AIDS, direct contact
6. allergy, immunity
7. lung cancer, disease
8. scurvy, vaccine

Workbook 24.6

1. I
2. N
3. I
4. N
5. N
6. I
7. I
8. N
9. I

Workbook 24.7

1. F
2. B
3. C
4. D
5. A
6. A
7. B
8. C
9. D
10. E

Workbook 24.8

1. Edward Jenner
2. used cow pox to vaccinate against smallpox
3. Joseph Lister
4. Robert Koch
5. Louis Pasteur
6. Rabies vaccine
7. Emil von Behring
8. 1890
9. Alexander Fleming
10. discovered penicillin
11. isolated vitamin C from lemons
12. 1935
13. first drug, sulfanilamide
14. polio vaccine
15. 1982

Workbook 24.9

1. F
2. B
3. P
4. G
5. I
6. H
7. E
8. M
9. C
10. D
11. A
12. N

Workbook 24.10

Natural immunity is inherited at the time of conception. Acquired immunity is produced during one's lifetime. Active acquired immunity is gained by coming into contact with an infectious disease and is long lasting. Passive immunity is given to a person by injections of antibodies and only lasts a short time.

Answer Key

Chapter 25

Workbook 25.2

1. U
2. I
3. R
4. L
5. I
6. C
7. R
8. L
9. I

Workbook 25.3

One possible answer: When certain **stimuli** affect a **receptor** they cause a **sensory neuron** to fire and a message is sent to the **spinal cord**. Here an **associative neuron** sends a message to a **motor neuron**. The motor neuron sends out an impulse on its **axon** and the **response** of the **effector** (muscle or gland) is an **unconditioned reflex**. The associative neuron also sends a message to the brain.

Workbook 25.4

zygote, embryo, fetus, baby, child, adult

Workbook 25.5

1. mitosis
2. meiosis
3. both processes
4. meiosis
5. mitosis
6. mitosis
7. meiosis

Workbook 25.6

1. unconditioned reflex
2. instinct
3. stimulus
4. learning
5. conjugation
6. gamete
7. sperm
8. egg
9. pure
10. hybrid
11. dominant

Workbook 25.7

1.

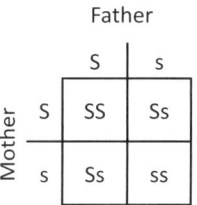

S = normal   s = sickle cells

Note to parent: The Punnett square does not predict the number of children the parents will have with sickle cell anemia. Rather, for each child, the Punnett square calculates the probability that he or she will have sickle cell anemia. Thus, even if the parents have already had three children without sickle cell anemia, the probability that a fourth child will have sickle cell anemia is still only 25%, not 100%.

2.

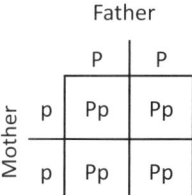

P = tasting   p = non-tasting

All of their children will be able to taste PTC.

3. 50%

Workbook 25.8

Formal Lab #7 Questions:

1. *Answers may vary.* Various experiments have shown that the brain responds more quickly to the sense of touch than to the senses of sight and hearing. Different people, however, may respond faster to different types of stimuli.

303

2. Answers will vary, but may include inaccuracy in measuring the distance; saying "Go" in the auditory test slightly before or after dropping the ruler, or at a slightly different time for different volunteers; variations in how close the volunteers held their hands to the ruler; the possibility that, by the third test, the volunteer may have learned how long the student usually waits before dropping the ruler.

Chapter 26

Workbook 26.2

A.

1. abiotic
2. abiotic
3. biotic
4. abiotic
5. biotic
6. biotic
7. biotic
8. abiotic
9. biotic
10. biotic

B.

1. food chain
2. niche
3. abiotic
4. producers
5. primary consumers
6. habitat
7. ecosystem
8. ecology
9. secondary consumers
10. biomes

Workbook 26.3

1. grass --> cow --> man
2. leaf --> leafhopper --> chicken
3. corn --> mouse --> snake
4. apple --> ants --> woodpecker
5. green algae --> tadpole --> bass

Workbook 26.4

One possible answer.

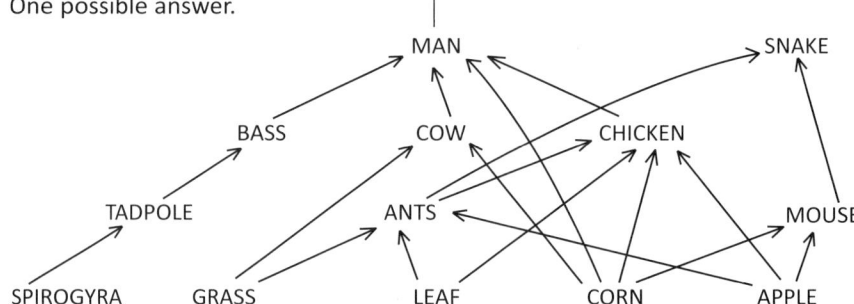

Workbook 26.5

A.

Tundra

　Plants: lichen or arctic poppy

　Animals: caribou or muskox

Boreal Forest

　Plants: pine tree, Venus flytrap, or sphagnum moss

　Animals: bald eagle, grizzly bear, or beaver

Tropical Rainforest

　Plants: orchid, rubber tree, or Venus flytrap

　Animals: boa constrictor or parrot

Desert

　Plants: saguaro cactus or tumbleweed

　Animals: sidewinding adder or roadrunner

Grassland

　Plants: prairie grass or tumbleweed

　Animals: antelope, lion, or buffalo

B. organism, habitat, ecosystem, biome, biosphere

Workbook 26.6

Compare the diagram to Figure 26.6 in the text. Check to make sure the diagram includes labels and a title.

Final Review

Part I. Keywords

The student should know the meanings and correct spellings of all the keywords.

Part II. Nutrition and Disease

1. $C_6H_{12}O_6 + 6O_2 \longrightarrow 6H_2O + 6CO_2 +$ Energy

2. $6CO_2 + 12H_2O \xrightarrow{\text{LIGHT ENERGY}} C_6H_{12}O_6 + 6O_2 + 6H_2O$

3. The five major nutrient groups are water, fats, proteins, carbohydrates, and vitamins and minerals.

4. See Figures 14.13 and 14.14 in the text.

5. To test for the presence of starch, place a drop of Lugol's solution, which contains iodine, onto the food sample being tested. If the Lugol's solution turns blue-black, starch is present.

6. **Deficiency disease:** any disease in Figure 24.3 in text.
   Environmental disease: lung cancer, skin cancer, diseases of the liver, etc.
   Functional disorder: any disease in Figure 24.4 in text.
   Genetic disease: color blindness, hemophilia, sickle cell anemia, Down syndrome, etc.

7. Infectious diseases can be transmitted by droplets, food, contact, cuts, and vectors.

8. Student should list two diseases from each category. Diseases described in the chapter are listed below.
   Viral diseases: smallpox, rabies (hydrophobia), yellow fever, polio, AIDS, the common cold, chicken pox, the flu, measles, warts, mumps, some cancers.
   Bacterial diseases: tuberculosis, diphtheria, pertussis (whooping cough), tetanus (lockjaw), typhoid fever, Lyme disease, scarlet fever.
   Protistan diseases: malaria, African sleeping sickness, amoebic dysentery.
   Fungal diseases: thrush, ringworm, athlete's foot.
   Animal diseases: tapeworms, hookworm, pork worm.

9. **Jenner** discovered a vaccine for smallpox.
   Pasteur discovered a vaccine for rabies.
   Koch discovered the bacterium that causes tuberculosis and developed Koch's Postulates for isolating disease-causing bacteria.
   Von Behring developed an antitoxin for diphtheria.
   Reed identified the yellow fever virus and its vector.
   Fleming discovered penicillin.
   Domagk discovered sulfanilamide, the first bacteria-killing drug.
   Salk made a vaccine for polio.

Part III. Body Systems

1. skeletal: irritability
   muscular: irritability
   digestive: nutrition
   circulatory: transport
   respiratory: respiration
   excretory: excretion
   endocrine: synthesis
   nervous: irritability
   reproductive: reproduction

2. hinge joint (ex: elbow, knees, fingers); gliding joint (ex: wrists, ankles, backbone); ball-and-socket joint (ex: hip, shoulder); immovable joint (ex: joints in the skull)

3. Voluntary muscles are muscles which you can control consciously. Involuntary muscles are muscles which your body controls without your conscious effort.

4. Two of the following:
   Amylase, found in the mouth and the small intestine, digests carbohydrates (starch).
   Pepsin, found in the stomach, digests proteins.
   Trypsin, found in the small intestine, digests proteins.
   Lipase, found in the small intestine, digests fats (lipids).
   Maltase, found in the small intestine, digests carbohydrates (maltose, a sugar).
   Sucrase, found in the small intestine, digests carbohydrates (sucrose, a sugar).

5. Plasma is the liquid part of the blood. Red blood cells are tiny, concave disks which carry oxygen to the cells. White blood cells engulf and destroy disease-causing microbes. Platelets are cell fragments which help in clotting the blood.

6. External respiration occurs in your lungs; cellular respiration occurs in each of your cells. In external respiration, oxygen in the air diffuses into your blood. In cellular respiration, the oxygen diffuses from your blood into your cells.

7. **Lungs:** carbon dioxide and water vapor
   Large intestine: solid wastes
   Urinary bladder: urine
   Skin: heat, water, salts, and nitrogenous wastes

8. *Student should identify four of the following glands, along with one of the hormones it secretes.*
   The **pituitary gland** secretes growth hormone to regulate growth, ACTH to regulate the adrenal glands, and TSH to regulate the thyroid gland.
   The **thyroid gland** secretes thyroxin to control the rate of metabolism.
   The **parathyroid glands** secretes parathormone to control the body's use of calcium and phosphate compounds.
   The **Islands of Langerhans** in the pancreas secrete glucagon and insulin to control the glucose levels in the body.
   The **adrenal glands** secrete adrenalin to help the body to respond in emergencies; it also secretes cortisol to regulate sugar metabolism and reduce inflammation and allergic reactions.
   The **male** and **female gonads** produce testosterone and estrogen which cause secondary male and female characteristics to develop.

9. The **cerebrum** controls thinking, sight, speech, and voluntary movement. The **cerebellum** controls coordination and balance. The **medulla**, or **brain stem**, controls the involuntary muscles.

Part IV. Animal Behavior, Reproduction, and Ecology

1. An unconditioned reflex is a simple, quick response to a stimulus which does not require the brain. A conditioned reflex is also a quick response to a stimulus, but this reflex involves the brain.

2. Mitosis and meiosis are types of cell divisions which both begin with a single cell, but mitosis results in two cells and meiosis results in four cells called gametes, which have only half the normal number of chromosomes.

3.
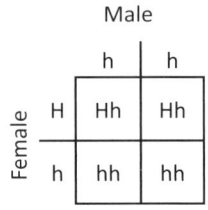

H = short hair   h = long hair

% short-haired: 50%

% long-haired: 50%

4. Answers will vary.

Part V. Diagram of Ear

Compare the diagram to Figure 23.13 in the text. Check to make sure the diagram includes labels and a title.

Part VI. Diagram of Carbon Cycle

Compare the diagram to Figure 26.5 in the text. Check to make sure the diagram includes labels and a title.